D1519314

Design of FET Frequency Multipliers and Harmonic Oscillators

For a complete listing of the *Artech House Microwave Library*,
turn to the back of this book.

Design of FET Frequency Multipliers and Harmonic Oscillators

Edmar Camargo

Artech House
Boston • London

Library of Congress Cataloging-in-Publication Data
Camargo, Edmar.
Design of FET frequency multipliers and harmonic oscillators / Edmar Camargo.
p. cm. — (Artech House microwave library)
Includes bibliographical references and index.
ISBN 0-89006-481-4 (alk. paper)
1. Microwave devices. 2. Frequency multipliers. 3. Millimeter wave devices. 4. Field-effect transistors. I. Title. II. Series.
TK7876.C36 1998
621.381'32363—dc21 98-33848
CIP

British Library Cataloguing in Publication Data
Camargo, Edmar.
Design of FET frequency multipliers and harmonic oscillators. — (Artech House microwave library)
1. Frequency multipliers 2. Microwave devices
I. Title
621.3'81'32'63

ISBN 0-89006-481-4

Cover design by Lynda Fishbourne

685 Canton Street
Norwood, MA 02062

International Standard Book Number: 0-89006-481-4
Library of Congress Catalog Card Number: 98-33848

10 9 8 7 6 5 4 3 2 1

To my wife, Marcia Rita
and
to my sons, Marcel and Regis

Contents

Preface

This book is intended as an introductory text on the application of field effect transistors (FETs) as frequency multipliers and harmonic oscillators, with the proper combination of theory and experimentation that will enhance learning and maintain interest, without extensive mathematics detracting from fundamental concepts and design rules. Moreover, the numerous illustrative design examples make this useful as a reference or supplementary book for any graduate course on nonlinear microwave design; or as a self-teaching text for students desiring to gain familiarity on large signal operation of microwave transistor circuits.

The area of Gallium Arsenide MEtal Semiconductor Field-Effect Transistor (GaAs MESFET) technology has undergone remarkable growth and development so that it became cost competitive to build functions with such devices that were traditionally performed by other devices like Schottky diodes, Varactor diodes, bipolar devices, etc., with considerable improvements on electrical performance. Therefore, besides employing transistors as amplifiers and oscillators, one will find in the literature frequency multipliers, frequency dividers, phase modulators, frequency mixers, and switches, just to name a few. In spite of the different objective of each of these applications, they all share a common property which is the application of the device in very large signal operations, exploring the device's nonlinearity to perform multiplication and generation of harmonics.

This book is mainly concerned with the techniques required to efficiently generate harmonics and emphasizes hardware circuits, not physics or mathematics. However, to properly follow the text, it is assumed there is a sound understanding of electronic circuit fundamentals, including diodes and FETs, a good knowledge of high frequency circuits, such as transmission lines, microwave amplifiers and oscillators, and graduate-level mathematics including Fourier analysis basics.

Following an introductory chapter on the application of frequency multipliers on telecommunication systems, and the devices used to perform such functions, nonlinear device modeling is introduced in Chapter 2 and the most used analytical approaches are analyzed and compared.

In Chapter 3, simple low frequency models are used to describe the multiplier operation. The reason to use such models is that basic graphic analysis can be employed to demonstrate current and voltage waveforms inside the device. This chapter is crucial if one is to perform the first fundamental step in multiplier design: selecting a transistor and determining the adequate bias. It also provides procedures for determining and extracting the basic parameters of frequency multipliers.

In Chapter 4, the nonlinear high-frequency models are considered, and the effect of parasitics on the gate and drain circuits are thoroughly discussed. The unavoidable feedback introduced by the parasitics or added to the circuit are also discussed and their effect on the multiplier performance is demonstrated. The circuit stability is included in this analysis and is an important design parameter. In this chapter the reactances are analyzed and their effect on the circuit design is demonstrated.

The design strategies for nonlinear circuits, including linearization, direct nonlinear synthesis, computer optimization, and harmonic load pull are introduced in Chapter 5. An example of an application of the linearization technique is presented with a significative importance: One can design a frequency doubler linearizing a MESFET dc model using only linear circuit analysis. The direct synthesis approach is also exemplified in the design of a frequency tripler.

Chapter 6 is devoted to harmonic oscillator design, where oscillator theory and frequency multiplier theory have to be joined in order to develop a component capable of performing a dual function. An application example demonstrates the efficacy of the design approach.

A large number of frequency multipliers topologies are given in Chapter 7, either to familiarize the reader with the options commonly found in the literature or to corroborate known topologies with simulated results. Most of the examples are for discrete components design but some examples of MMIC designs with commercially available MMICs are shown. A few examples of harmonic oscillators complete the chapter.

The content of this book is an outgrowth of my experience on nonlinear design that started at the Centre National D'Etudes des Télécommunications (CNET), located in Lannion, France, in the early 1980s. It also represents a recollection of lecture notes taught by the author in graduate courses at the University of São Paulo, Brazil, and the results on research work developed at

Laboratório de Microeletrônica (LME) at the same university. It represents the interchange of knowledge among my students, contributors from France, Dr. Robert Soares, Dr. Michel Goloubkoff, my former mentor-teacher from Brazil, Dr. Jose Kleber da Cunha Pinto, Dr. Fatima S. Correra at the University of São Paulo, and Dr. Amarpal S. Khanna at Hewlett Packard in Santa Clara, California. I remain indebted to all of them for the direct or indirect help they gave me in the preparation of this book.

A special acknowledgment is due to Dr. R. A. Perichon from University of Brest and my advisor at CNET. During the preparation of this book he was a serious and constructive critic who gave important suggestions that enriched this work. Another engineer that deserves to be acknowledged is Luiz Antonio Razera for his contribution in the realization of an adequate manuscript. I can't see how this work could have been completed without his help.

Edmar Camargo
San Jose, California
September, 1998

1

Introduction

A key element in radio systems is the availability of a "clean" local oscillator. The word "clean" means a signal free from any sort of spurious frequencies generated by natural and man-made sources. Most noises are generated external to the signal source and are eliminated through a correct filtering. The remaining sources of noise are due to the oscillator active device and the techniques used to design the local oscillator. The quality of a signal generator can be evaluated according to either its long-term or short-term stability. Long-term stability refers to changes in the frequency over a reasonable time duration. The frequency drift observed for a few minutes due to thermal effects is a good instance of long-term stability. Another example of such stability is the frequency drift due to circuit aging, an important phenomena in crystal oscillators. Short-term stability refers to low frequency or phase modulation of the generator. They usually result from random phenomena, are also known as "phase noise" of a generator, and correspond in the frequency domain to the sidebands located above and below the main signal centered at frequency f_0, depicted in Figure 1.1(a).

This type of noise is important in radio systems since it cannot be "filtered." It may eventually be down-converted, and fall into the intermediate frequency (IF) band, degrading system noise performance. This degeneration is even more critical in digital systems because the down-converted noise is detected in the demodulator generating jitter in the recovered digital signal, shown in Figure 1.1(b). The toleration of such interfering noise depends on the modulation scheme. The most common digital modulation schemes employed in digital radios are binary phase shift keying (BPSK), quadrature phase shift keying (QPSK), and m-quadrature amplitude modulation (mQAM) where *m* stands for the number of symbols, specifying how the carrier phase

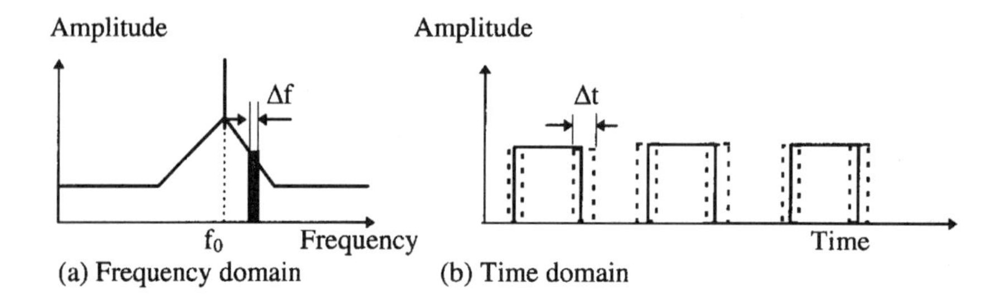

Figure 1.1 Phase noise on a local oscillator.

and amplitude are modulated. Each of these systems is plotted on an I-Q diagram (Figure 1.2). In this diagram, each point is the result of vector addition of I, in-phase and Q, quadrature-phase components. In the first two, only the phase is modulated while in the third, both phase and amplitude are modulated. In principle, as long as the decision boundaries are respected, the jitter is acceptable. Figure 1.2 shows the diagram of simulations on a demodulated BPSK, QPSK and 16 QAM signals, whose source [1] was subjected to a phase modulation of 5.53 degress. Other factors such as spurious frequencies, thermal noise, and AM-PM conversion are not considered in this example.

From this set of figures it is obvious that a BPSK-modulated signal can tolerate more jitter than a 16 QAM one. In the latter it is evident that boundaries are being crossed, resulting in inter-symbol interference and an increase in the probability of error during transmission, which is usually specified in terms of bit error rate (BER). For this reason, digital communications equipment specifications are very demanding on low-phase noise. They are specified in terms of dBc/Hz related to an offset frequency from the carrier. A summary [2] of the noise specs for mm-Wave radios for the most common modulation schemes are listed in Table 1.1. The requirements for more complex modulation

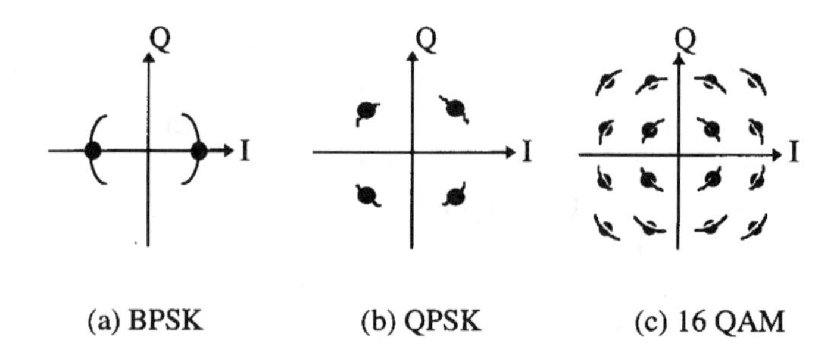

Figure 1.2 Jitter on a demodulated digital signal. (Reprinted with permission from *Microwave Journal,* July 1977).

Table 1.1
Phase Noise Specifications for Digital Radios

BPSK	–85 dBc/Hz @ 100 kHz
QPSK	–90 dBc/Hz @ 100 kHz
16QAM	–90 dBc/Hz @ 10 kHz

like 16 QAM are so stringent that phase noise has to be specified closer to the carrier.

Obtaining these specifications in mm-Wave frequencies with a fundamental oscillator is a challenge if system cost tradeoffs are made. Although it may be technically possible to build a fixed frequency cavity-tuned generator that provides the above specifications, it will not be accepted by radio systems where agile sources are required to ease implementation of radio links. Therefore, the traditional solution of generating a low frequency low noise source followed by frequency multiplication, as illustrated in Figure 1.3, is a preferred solution by radio manufacturers. In spite of the phase noise degradation given by the relation $10\log_{10}N$, where N is the multiplication order, this solution gives a good compromise between cost and final phase noise in many radio systems. However, this degradation may preclude the use of frequency multipliers, especially in the case of high multiplication order.

The advantage of such a scheme is clarified by Figure 1.4, where a fundamental VCO is compared to a VCO followed by a times 4 multiplier. It can be observed that mm-Wave fundamental oscillators perform poorly mainly for two reasons: (1) degradation of the Q-factor of the tank circuit at those frequencies; and (2) the technology employed in high-frequency devices causes high $1/f$ noise, contributing to the degradation of its performance. This

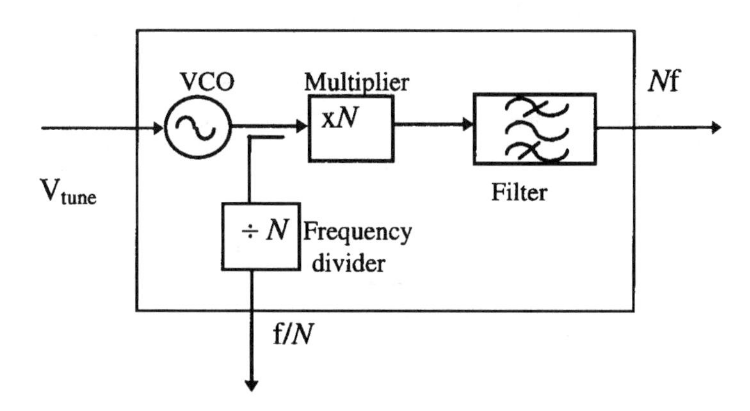

Figure 1.3 Generation of a clean signal.

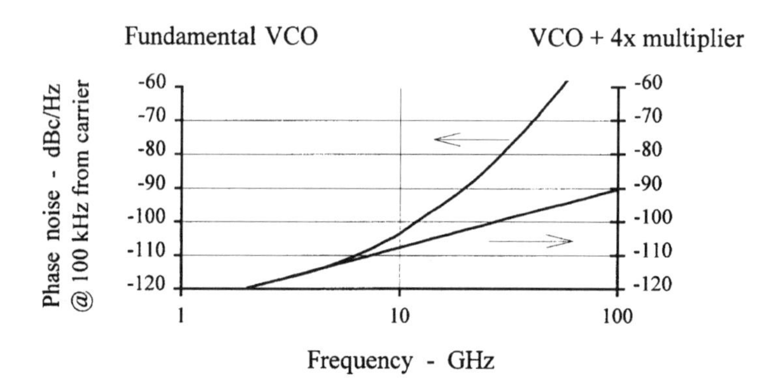

Figure 1.4 Phase noise: fundamental source and multiplied source.

trend is also valid for multipliers of different orders, where the phase line slope will change accordingly.

Another interesting system application of a multiplier is the derivation of a local oscillator coherent with a fundamental one, as exemplified in Figure 1.5. This figure shows the up-conversion of a modulated carrier to mm-Wave by use of a dual-conversion approach. The advantage of such a system is the use of a higher IF frequency at the second conversion easing the task of building very sharp filters and avoiding excessive leakage of the LO signal at the output. That is a cost effective solution since only one oscillator is needed, and the second is generated by a multiplier which is a much simpler circuit.

There are two conventional approaches to generate a frequency multiplier: harmonic generation by nonlinear conductance, and harmonic generation by nonlinear reactance. The former employs a Schottky-barrier diode to create a half-wave sinusoid that is rich in even harmonics. The latter uses the nonlinear capacitance of a varactor diode to generate harmonics, which can also be used

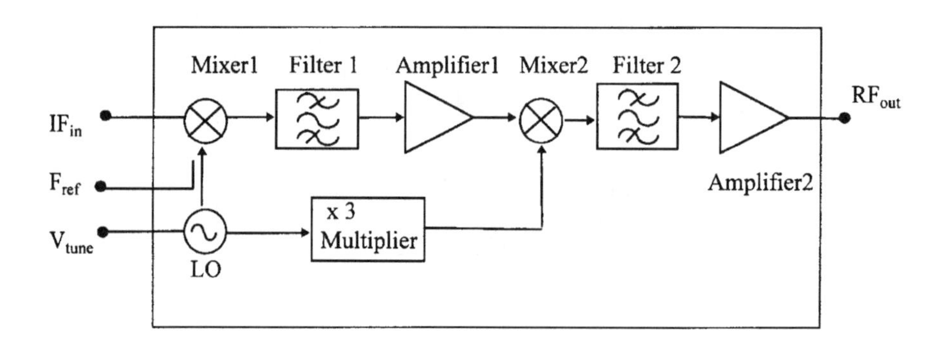

Figure 1.5 Use of a multiplier in a dual up-conversion system.

as an up-converter, or as a parametric amplifier. These multipliers are typically highly efficient due to low losses, and are capable of providing power in the order of watts. However, they generate significantly more phase noise [3] than multipliers based on nonlinear conductance. Higher amplitude modulation (AM) to phase modulation (PM) conversion efficiency, as well as charge fluctuations produced by changes in minority carrier lifetimes, are suspected to cause this difference. Schottky diodes have low flicker noise and essentially no minority carrier storage, and are therefore more adequate to low-phase noise frequency multiplication.

1.1 Phase Noise Fundamentals

The signal from a generator can be the result of amplitude modulation and frequency or phase modulation of a carrier by the noise vectors V_{AM} and V_{PM}, which are represented in Figure 1.6. The carrier is a pure sinusoid of fixed amplitude, frequency and phase, while the noise vectors are of much lower amplitude, and their magnitude and phase vary randomly.

The phase modulation originates from the low frequency noise of the oscillator active device. At these frequencies, the noise vector power is inversely proportional to frequency: close to dc the noise decreases proportional to $1/f^3$, an effect known as *frequency modulated flicker noise*; then it decreases proportional to $1/f^2$, due to frequency modulated white noise. This type of noise continues to decrease and crosses frequency, f_1, denominated corner frequency, where the AM noise level is greater than the FM level and is constant over frequency. The corner frequency is in the range of kHz, and above that frequency the FM noise is negligible compared to the AM noise. Therefore, the oscillator phase noise can be explained as a result of the nonlinear nature of oscillators, which up-converts the low frequency noise, resulting in the two side bands close to the carrier. This effect is illustrated in Figure 1.7 where the AM noise is not multiplied but added to the oscillator spectrum.

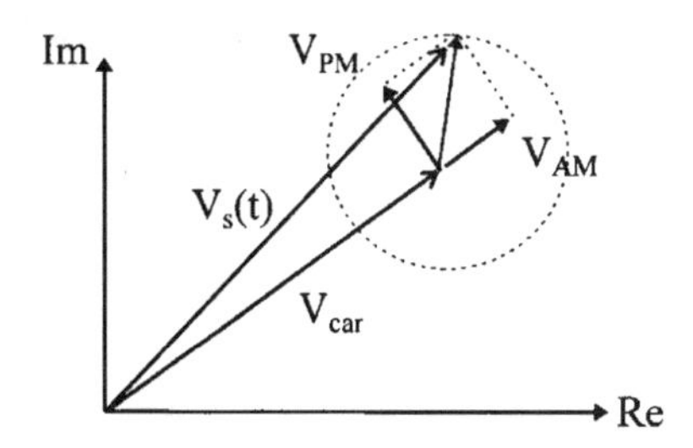

Figure 1.6 Amplitude and phase modulation of a carrier.

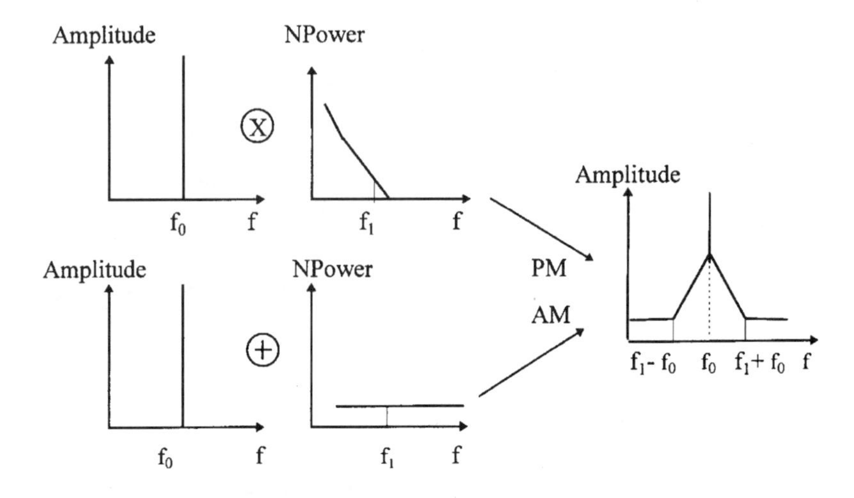

Figure 1.7 Phase noise as a result of up-conversion.

A simple mathematical treatment [4] to study noise in oscillators can be obtained from the amplitude and phase modulation of a carrier described by

$$V_s(t) = V_{\text{car}}[1 + m(t)]\cos[2\pi f_c t + \phi_p(t)] \tag{1.1}$$

where,

$m(t)$ = amplitude modulation
$\phi_p(t)$ = phase modulation

Neglecting the amplitude modulation, the frequency modulation of a signal carrier, f_c, by a sinusoidal of frequency f_m, is defined by (1.2), and the frequency-phase relation is given by the integration of frequency in time:

$$f(t) = 2\pi f_c + 2\pi\Delta f_{\text{peak}}\cos(2\pi f_m t) \tag{1.2}$$

$$\phi_p(t) = \int 2\pi f(t)dt \tag{1.3}$$

Performing the integration and inserting into (1.1), results in

$$V_s(t) = V_c\cos[2\pi f_c t + (\Delta f_{\text{peak}}/f_m)\sin(2\pi f_m t)] \tag{1.4}$$

where,

f_c = signal carrier frequency

f_m = modulating frequency or offset frequency from carrier

$V_s(t)$ = amplitude of the signal

$\phi_p(t)$ = instantaneous angle modulation

Δf_{peak} = maximum frequency deviation

$\Delta f_{peak}/f_m = \beta_m$ = frequency modulation index

Equation 1.4 may be written in the form

$$V_s(t) = V_{car}\cos(2\pi f_c t)\cos[\beta\sin(2\pi f_m t)] - V_{car}\sin(2\pi f_c t)\sin[\beta\sin(2\pi f_m t)] \tag{1.5}$$

Now the $\cos(\beta\sin(2\pi f_m t))$ and $\sin(\beta\cos(2\pi f_m t))$ functions may be expanded as Fourier series whose coefficients are ordinary Bessel functions of the first kind with argument β. These functions are described by

$$\cos[\beta\sin(2\pi f_m t)] = J_o(\beta) + \sum_{n=\text{even}} 2J_n(\beta)\cos(n2\pi f_m t) \tag{1.6}$$

$$\sin[\beta\sin(2\pi f_m t)] = \sum_{n=\text{odd}} 2J_n(\beta)\sin(n2\pi f_m t) \tag{1.7}$$

Substituting (1.6) and (1.7) into (1.5) and expanding, the products of sines and cosines results

$$\begin{aligned} V_s(t) = {} & V_{car}J_o(\beta)\cos(2\pi f_c t) + \\ & \sum_{n=\text{odd}} V_{car}J_n(\beta)[\cos(2\pi(f_c + nf_m)t) - \cos(2\pi(f_c - nf_m)t] + \\ & \sum_{n=\text{even}} V_{car}J_n(\beta)[\cos(2\pi(f_c + nf_m)t) + \cos(2\pi(f_c - nf_m)t] \end{aligned} \tag{1.8}$$

The phase noise corresponds to a carrier frequency modulated by a single tone frequency, f_m, with a very low modulation index, $\beta \ll 1$. Therefore, only the first two terms of (1.8) need to be considered, so that the resulting spectrum consist of the carrier and two line spectra located at frequencies $(f_c + f_m)$ and $(f_c - f_m)$, as shown in Figure 1.8. Under these conditions, it is demonstrated that the following approximations can be made

$$\begin{aligned} J_o(\beta) &\approx 1 \\ J_1(\beta) &\approx 0.5\beta = 0.5(\Delta f_{peak}/f_m) \end{aligned} \tag{1.9}$$

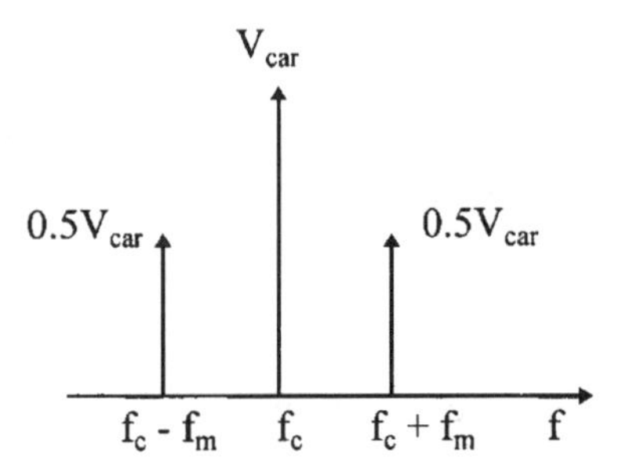

Figure 1.8 Phase noise spectrum.

The single side band phase-noise to carrier ratio is obtained by dividing the upconverted voltage at either $(f_c + f_m)$ or $(f_c - f_m)$ by the carrier voltage, as shown in (1.10),

$$V_{SSB}/V_s = 0.5(\Delta f_{\text{peak}}/f_m) \tag{1.10}$$

where, V_{SSB} corresponds to the amplitude of the single-sideband noise.

Extending the frequency components of the modulating signal from zero frequency up to the f_m frequency, the modulating signal becomes noise, whose power density is described by (1.11). It is normal practice to square that expression to transform the voltage ratio into a power ratio and express it in terms of decibels:

$$\begin{aligned} \mathscr{L}(f_m) &= 10\log(V_{SSB}/V_s)^2 = 20\log 0.5(\Delta f_{\text{peak}}/f_m) \\ &= -6 \text{ dB} + 20\log(\Delta f_{\text{peak}}/f_m) \end{aligned} \tag{1.11}$$

The phase noise power at a distance f_m from the carrier is depicted in Figure 1.9 and is measured in a 1 Hz bandwidth.

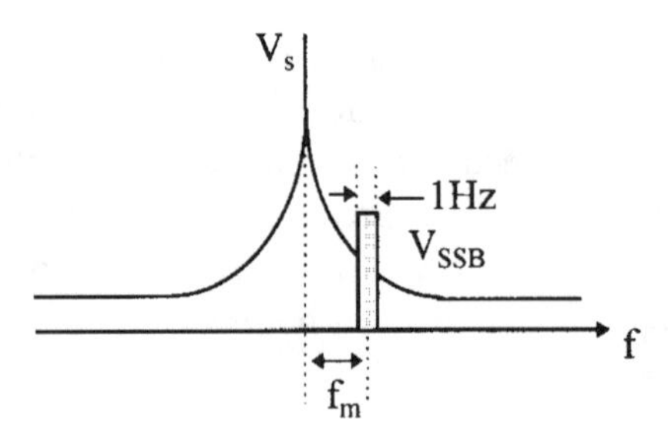

Figure 1.9 Phase noise power.

The convenient form of (1.11) allows the increase in phase noise to be calculated when the frequency of a signal is multiplied. When a signal, $f \pm \Delta f$, is multiplied by a factor N, the frequency deviation is multiplied by the same amount, but the rate of modulation remains the same. Therefore, relating (1.10) for the original Δf_{peak}, and the multiplied frequency deviation, $N\Delta f_{\text{peak}}$, one obtains:

$$\mathscr{L}(Nf)/\mathscr{L}(f) = 20\log(Nf/f) = 20\log(N) \tag{1.12}$$

A more complete treatment [5] of noise degradation in frequency multipliers is given by the noise transmission matrix, (1.13). This equation is valid if the noise is Gaussian and is stationary.

$$\begin{bmatrix} \beta_{\text{out}} \\ m_{\text{out}} \end{bmatrix} = \begin{bmatrix} T_{pp} & T_{pa} \\ T_{ap} & T_{aa} \end{bmatrix} \cdot \begin{bmatrix} \beta_{\text{in}} \\ m_{\text{in}} \end{bmatrix} \tag{1.13}$$

The coefficients m_i, β_j represent amplitude and phase modulations, and the AM and PM noise present in the signal applied to the multiplier are represented by m_{in} and β_{in}, respectively. The transmission matrix indicates the conversion coefficients (PM to PM, AM to AM, AM to PM, PM to AM) at a given modulation frequency. In an ideal multiplier, $T_{pp} = N$, $T_{aa} = 1$ and $T_{ap} = T_{pa} = 0$. Thus, the phase noise is degraded by the multiplication factor N, and all other levels remain constant. Experimental measurements in FET frequency multipliers show that this is a good description of noise degradation for frequency doublers. For higher order multipliers it has been observed that $T_{pa} \neq 0$, so that there is an additive phase noise degradation due to AM to PM conversion, which increases when the active device is driven into deep saturation. While the phase noise is a concern at frequencies close to the carrier, the AM noise is important at frequencies far away from the carrier. In real circuits, T_{ap} is also proportional to N, and the factor T_{aa} represents the gain compression characteristics of the multiplier. Thus, amplitude noise is also degraded when processed by a multiplier. If the multiplied signal is amplified, then the noise power from the amplifier is added to the multiplier noise. Therefore, a multiplied source presents a performance opposite to that described by Figure 1.4 with respect to noise far from the carrier: The AM noise is higher compared to fundamental oscillators at the same frequency. In communication systems the amplitude degradation is not so important, since this type of noise can be filtered before being applied to a mixer. Its effect is further minimized if balanced mixers are employed.

A low phase noise source is dependent on the following key parameters: the choice of the active device, the choice of the resonator, and the method by which they are coupled together. The choices on active devices are many:

- Silicon bipolar junction transistors;
- Heterojunction bipolar transistors;
- HBTs;
- The FET family: GaAs MESFETs, HEMTs (high electron mobility transistor), and PHEMTs (pseudomorphic HEMT);
- Gunn diodes.

The bipolar devices and Gunn diodes are the ones offering lower $1/f$ noise, thus, the better choices for low phase noise. Gunn diode oscillators have traditionally been reasonably quiet sources at high frequencies, but they suffer from microphonics, temperature effects, controllability, and other factors. FET devices have been successfully employed in the design of high frequency fundamental oscillators, but they are quite noisy even when coupled to very high "Q" cavities. Silicon BJTs are the best devices for oscillator application, but their frequency of operation is limited to the top of X-band. This problem is overcome by cascading them with frequency multipliers.

Low-phase noise oscillators use a high Q-factor resonator, or tank circuit, whose coupling to the active device circuit is designed in such a way that the Q-factor degradation is minimal. The availability of high dielectric constant material and its application as a tank circuit led to the development of dielectric resonator oscillators (DROs). They present low phase noise at low cost, but their tunability is limited which restrict their application in digital radios. Conventional LC resonators built with varactors present octave bandwidth tuning with a reasonable phase noise. Usually such sources are employed when phase noise in the order of –85 dBc/Hz @ fm = 100 kHz is acceptable. A ferromagnetic resonator built with a Ytrium iron garnet (YIG) crystal possesses a high unloaded Q, can be tuned over a wide range of frequencies, and presents noise levels of –80 to –90 dBc/Hz @ f_m = 10 kHz. They are currently tuned by means of a variable magnetic field controlled by a current which makes them an expensive choice compared to other options.

1.2 Schottky-Barrier Diode Multipliers

The current through a Schottky-barrier diode is given by (1.13), which shows an exponential relation between applied voltage and current flowing through the diode.

$$I = I_s[\exp(x) - 1] \tag{1.14}$$

where,

$x = qV(t)/kT_a = (qV_s/kT_a)\cos(\omega t)$

q = electron charge

k = Boltzman constant

T_a = absolute temperature

I_s = diode saturation current

This equation can be expanded into a Fourier series expansion given by (1.14), where $I_n(x)$ is the modified Bessel function of the first kind [6].

$$I = I_s[I_0(x) + 2\sum_{1}^{\infty} I_n(x)\cos(n\omega t)] \tag{1.15}$$

A plot of the harmonic current as a function of applied voltage, normalized by its dc component, is shown in Figure 1.10. The difference between the harmonic levels is higher at low applied voltages. Increasing the applied voltage, the second and third harmonic components also increase and their amplitudes tend to saturate to a constant value.

A half-wave rectifier doubler circuit is obtained by applying a relatively large amplitude sinusoidal voltage of period 2π, to the diode plus load circuit represented in Figure 1.11(a). During the positive semicycle, the applied voltage is greater than the thermal voltage (q/kT_a) and the diode conducts current to the load, R_L. The generator voltage is applied across the diode and load resistor, at the generator frequency. On the negative semi-cycle, the voltage across the diode is lower than the thermal voltage and the exponential term of (1.13) becomes negligible. There is no current flowing in the circuit. The current

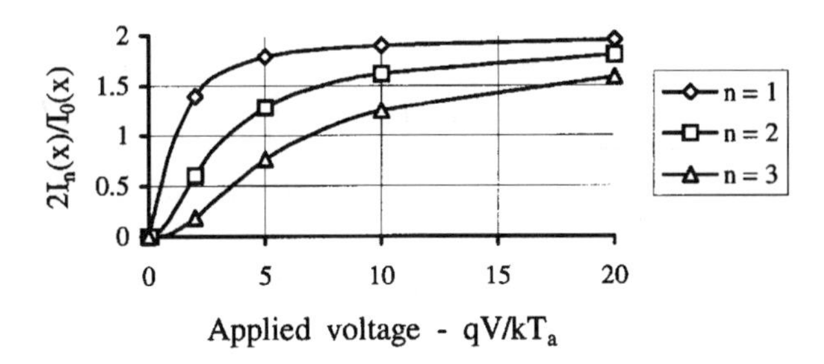

Figure 1.10 Bessel function ratios versus applied voltage.

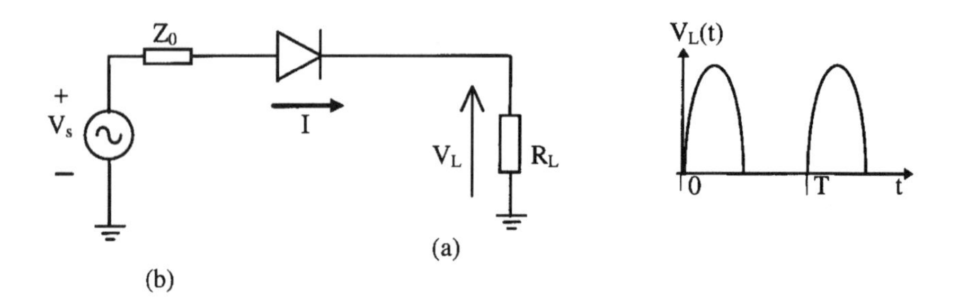

Figure 1.11 Single diode multiplier (a), and load voltage (b).

applied to the load generates a half-wave sinusoid voltage waveform, shown in Figure 1.11(b), rich in even harmonics. The second harmonic can be selected by adequate filtering.

A better multiplication efficiency can be obtained by antiparalleling two diodes by means of a 180 degree hybrid as shown in Figure 1.12(a). In this case, the applied voltage is split in two paths with a phase difference of 180 degrees between them. In the positive semicycle of the applied sinusoid, the diode connected to the 0 degree path conducts delivering current to the load. On the negative semicycle, the voltage on the other path is positive, due to phase shift introduced by the hybrid, causing conduction of the other diode. Thus, both cycles of the sinusoid deliver current to the load instead of just one cycle. The load current, given by (1.15) and (1.16), is the sum of the currents through each diode. The load voltage shown in Figure 1.12(b), does not contain the fundamental and odd harmonics, but shows a dc level on the output.

$$I_{\text{total}} = I_1 + I_2 = 2I_s[\cosh(x) - 1] \tag{1.16}$$

$$I_{\text{total}} = 2I_s[I_0(x) - 1 + 2\sum_{1}^{\infty} I_n(x)\cos(2n\omega t)] \tag{1.17}$$

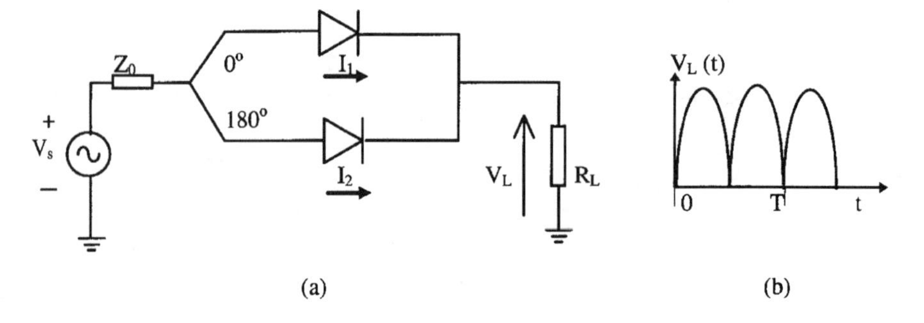

Figure 1.12 Balanced diode multiplier (a), and load voltage (b).

The generation of odd harmonics requires paralleling the diodes as shown in Figure 1.13(a). In this case, one diode conducts during the positive semicycle while the other conducts during the other semicycle, so that the current in the load is given by the difference of current in each diode.

$$I_{\text{total}} = I_1 - I_2 = 2I_s[\sinh(x)] \tag{1.18}$$

$$I_{\text{total}} = 4I_s\left\{\sum_{0}^{\infty} I(2n+1)(x)\cos[(2n+1)\omega t]\right\} \tag{1.19}$$

One can conclude from (1.18) that there is no dc component, but the fundamental component and odd harmonics are present in the total current. If the applied voltage is high enough $(\gg q/kT_a)$, there results a symmetrical distortion in the waveform given by (1.18), resembling a square waveform. Note this configuration requires a bandpass filter to remove the fundamental component and higher harmonics.

1.3 MESFET Multipliers

This type of device operates in the linear mode as long as the terminal voltages are contained within the limits described in Chapter 3. Outside those limits the device is highly nonlinear, therefore adequate to be employed in mixers and in frequency multipliers. Unlike diodes, MESFET's currents cannot be described by a single equation and thus require a more complex mathematical treatment to describe their nonlinear operation. Comparison of MESFET and Schottky diode multipliers is not simple and requires engineering judgment for the specific application. For instance, frequency doublers built with Schottky-barrier diodes, are lossy (in general > 10 dB), require no tuning elements, and can operate over octave bandwidth, a performance that cannot be attainable

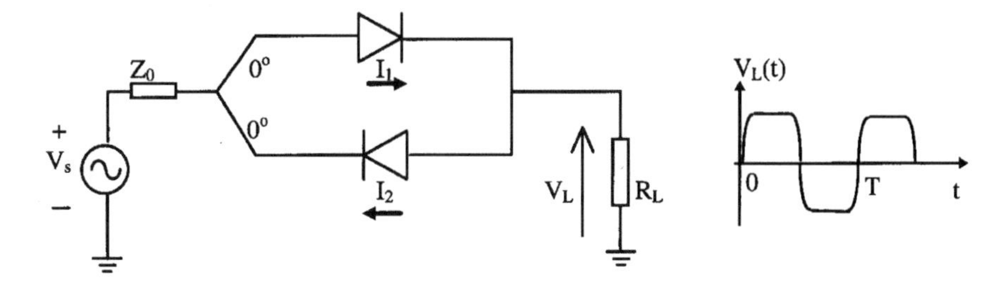

Figure 1.13 In-phase diode multiplier (a), and load voltage (b).

with FETs. On the other hand, if bandwidth is less than 50%, FETs present lower losses and eventually can provide gain depending on bandwidth, frequency, type of device, and other characteristics that will be addressed in this book. One can, however, use intuition to conclude that transistors offer several other advantages compared to diodes: isolation from input to output so that usually it suffices to match the input at the fundamental frequency, and the use of only one bandpass filter at the output. They also require much lower drive power for generating a significative output power level. If higher order multipliers are considered (x3, x4, etc.), then Schottky-barrier diodes perform poorly and cannot compete with FETs. An additional advantage is that FETs are a natural part of MMICs where they can be designed to fit a specific need. This book discusses the design and application of single-gate GaAs FET devices to frequency multipliers. Other options like the dual-gate will not be considered, but the conclusions of this book are equally applicable. A dual-gate device can be considered as the cascade of two devices, where one operates as a frequency multiplier and the other as an amplifier.

References

[1] Simpson, M. R., and J. M. Dixon, "The Application of Low Noise, X-Band Synthesizers to QAM Radios," *Microwave Journal,* July 1977, pp. 76–85.

[2] Khanna, A. P. S., "Microwave/mm-Wave Components & Technology for Short Haul Digital Radio Links," Asia-Pacific Microwave Conference Proceedings, 1996.

[3] Hoberg, L., "Multiplier Circuits Keep Phase Clean," *Microwaves,* June 1982, pp. 71–81.

[4] Carlson, B., *Communication Systems,* New York: McGraw-Hill Book Company, 1968.

[5] Riddle, A., "A CAD Program and Equations for System Phase and Amplitude Noise Analysis," IEEE MTT-S 1989 International Microwave Symposium Digest, Long Beach, CA, 1989, pp. 359–362.

[6] Clarke, K. L., and D. T. Hess, *Communication Circuits: Analysis and Design,* Florida: Krieger Publishing Company, 1994.

2

Nonlinear MESFET Models

The fundamental hypothesis which governs the nonlinear FET models described in this chapter is the assumption that the large-signal performance of microwave transistors is essentially governed by its dc characteristics [1]. This quasi-static assumption has been verified experimentally by several authors [1, 3] to be valid as long as the frequency of operation is such that the model can be represented by time-invariant lumped elements. The elements on this type of model have a strong dependence on dc bias and on the amplitude of the terminal dynamic voltages. They usually require extensive measurements to yield an accurate model. Without such models, one must rely on experience, and on bench trial and error experiments to extract the desired device performance. A nonlinear model provides insight into the circuit operation, on the interaction of device parameters with external circuit impedances within a large frequency band, and on the current versus voltage relationship, along with other important information leading to an optimum design. This information is otherwise unavailable to the circuit designer. However, it must be emphasized that a general purpose nonlinear equivalent model capable of simulating any nonlinear condition still remains to be developed. Therefore, each one of the models discussed in this chapter have limitations.

In this context, the proposed nonlinear models were constructed from a classical small-signal model, assuming it is a first-order approximation of a circuit capable of representing the device nonlinear performance. Thus, the nonlinear elements which are dependent on the external bias voltages can be reduced to the small-signal model at any bias point. The analytical models presented here can be determined from static measurements and from small-signal measurements. Their accuracy is acceptable to simulate the generation of harmonics from a MESFET model. The word acceptable has to consider the technology employed in the circuit construction. In hybrid microwave integrated circuits (MICs), the requirements for accuracy are not so stringent since the final circuit can be tuned to achieve the desired performance. In monolithic microwave integrated circuits (MMICs), the model of accuracy is much more important since tunability in this technology is very limited and design iterations are very expensive. The design philosophy is targeted to a final circuit that works the first time it is built. In this case, the models presented herein can still be used but they will require more time and sophisticated equipment to improve their accuracy.

2.1 The Static FET I–V Characteristic

The cross-section of an *n*-type Schottky-barrier gate MESFET is represented in Figure 2.1. It consists of an active layer grown on top of an insulating GaAs substrate which serve as a mechanical support for the structure. The layer under the gate is called the channel and is uniformly doped with donors, which will carry the current from the drain to the source terminals. The layers under the source and drain are made ohmic with a low resistance to improve the device performance.

The gate metal on top of the active region is the anode of the Schottky-barrier diode and the conducting top layer of the GaAs is the cathode. Applying a negative bias to the gate with respect to the channel, electrons are pushed away, reducing the thickness of the layer which can carry current. If it is negative enough, the channel is completely depleted of electrons and there is no current flow through the channel. If the gate and source are grounded and the drain voltage is increased from zero to V_{DM}, the current initially increases proportionally to the voltage until it reaches the knee voltage, V_k, depicted in Figure 2.2. That voltage separates the device resistive region where there is a linear relation between voltage and current, from its active region ($V_{DS} > V_k$) where the drain current is approximately constant.

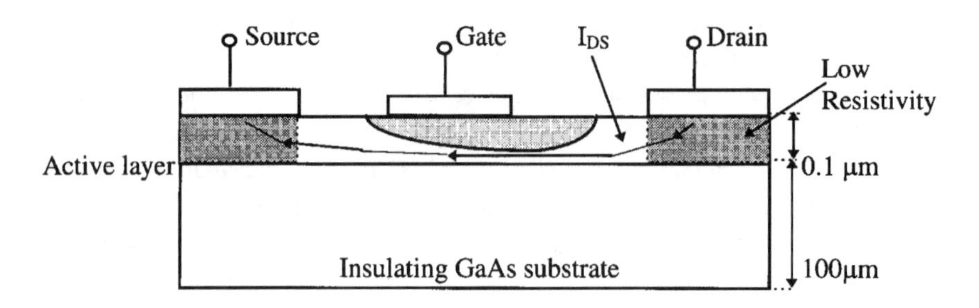

Figure 2.1 Cross section of a typical MESFET.

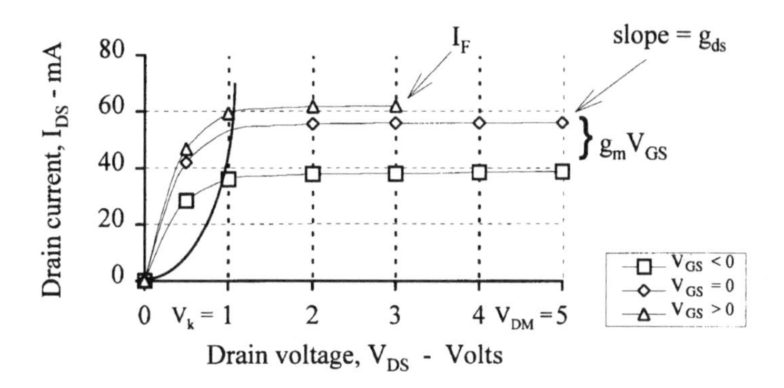

Figure 2.2 $I_{DS} - V_{DS}$ output characteristics.

Observe that this boundary will change for different drain currents, but V_k is defined specifically for the condition V_{GS} = 0. Increasing the voltage further, the drain current remains almost constant and the small changes observed are due to the drain-source output conductance. The current saturates because the electron mobility inside the channel reaches a maximum velocity after a certain electrical field magnitude has been exceeded. This velocity is a property of material and it is around 1×10^7 cm/s for GaAs. The drain current for this gate condition, V_{GS} = 0, is called the *saturated drain current*, I_{DSS}. If the drain voltage continues to increase, it will eventually cross the gate-drain barrier breakdown voltage, BV_{GD}, enter in avalanche, and a catastrophic failure will occur.

The dc transfer characteristic (transfer of input voltage to output current) in Figure 2.3 is for a drain voltage of 3.0V. At pinch-off, V_P, the channel is completely depleted and there is no drain current. In reality, V_P is parameter defined for JFET's and is often confused with the threshold voltage, V_T, which

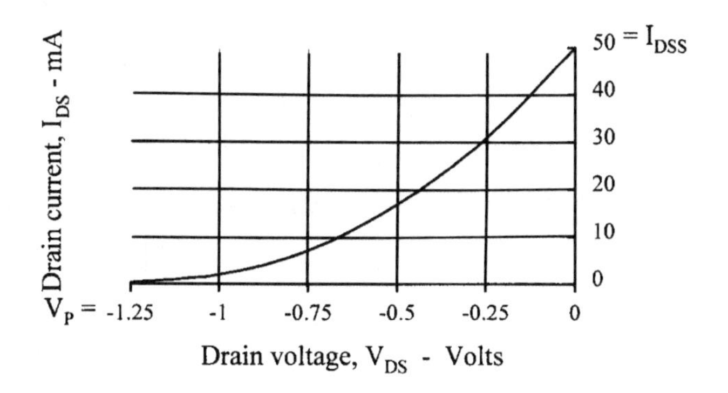

Figure 2.3 $I_{DS} - V_{GS}$ transfer characteristics.

is the gate voltage that depletes the channel. The relation between threshold and pinch-off is given by $V_P = V_T + V_\phi$. Throughout the book, pinch-off voltage will be used as defined in this page. Increasing the gate voltage, the current increases in proportion to the dc transconductance and at $V_{GS} = 0$ volts, reaches the current defined as I_{DSS}.

Increasing the gate voltage further, the current continues to increase up to the point where the gate-source diode starts conducting. At that point, the device reaches the maximum drain current, I_F, which occurs when the gate voltage is around 0.7V. The drain current does not follow the gate voltage beyond this point and if the voltage is too high the device will start degrading and will eventually be destroyed.

2.2 Model Description

Adding one dimension to the MESFET cross-section of Figure 2.1, one obtains a complete description of the device structure, depicted in Figure 2.4. That description is very useful to identify the elements of the *nonlinear model.* The Schottky-barrier from gate-to-source and gate-to-drain gives origin to two diodes D_{GS} and D_{GD}, respectively. The capacitances C_{gs} and C_{gd} express the changes in the depletion layer charge as a function of two internal voltages, V_{di} and V_{gi}. The resistance R_i is associated with the time constant required to charge the C_{gs} capacitor. The effect of controlling the thickness of the channel by controlling the gate voltage is represented by a current source in

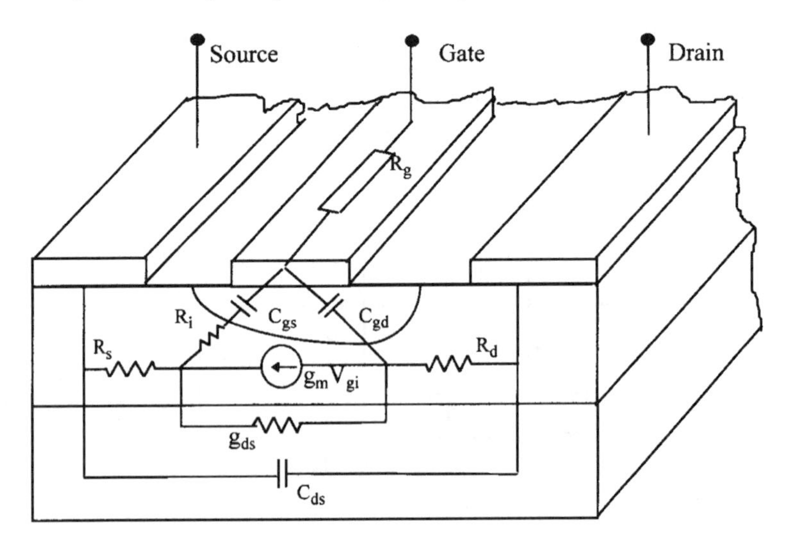

Figure 2.4 Physical origins of model elements.

the channel, I_{DS}, which is controlled by V_{gi} and V_{di}. The linear parasitic elements include the drain-to-source capacitance, C_{ds}, due to the bonding pad area, and the resistances: the ohmic-contact series resistances R_s and R_d which are in series with the source and drain; the gate resistance due to the gate metalization which is in series with the gate, R_g.

The Schottky-barrier diodes usually do not participate in the device operation since they are reverse biased. At large dynamic voltages there may be forward conduction or breakdown currents for a small fraction of the signal period. If they are limited in amplitude, it is possible to operate the device in this condition without degrading reliability. This instantaneous conduction has an important effect on the device's overall performance, a good reason why the diodes have to be part of the model. The nonlinear current source I_{DS} is the most important nonlinear element since it is responsible for power gain in an amplifier and generation of harmonics in a frequency multiplier.

The model represented in Figure 2.5 is quite general, adequate for nodal analysis, and can be applied to different types of field-effect devices such as MESFETs, HEMTs and PHEMTs. A good description of HEMT and PHEMT operation is found in the literature [2]. They present higher transconductance, operate at higher frequencies, and their physical structure is more complex when compared to MESFETs. However, for the purposes of this book it is assumed that their nonlinear characteristics are similar to MESFETs.

2.3 Analytical Models

These models use an analytic equation to describe the nonlinear current source as a function of terminal voltages, are simple to use, and provide results with

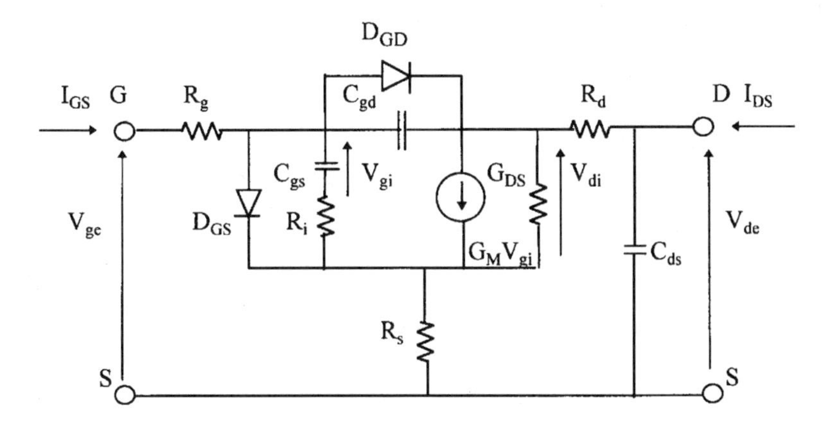

Figure 2.5 Nonlinear equivalent model.

a good degree of accuracy. In general, an analytic equation is also developed for the capacitances as a function of voltage and are incorporated into the model. There are several models available in the literature of varying capabilities to predict the performance of FETs in nonlinear operating modes. This area is still under development, with new approaches being introduced every year. In this context, a few models were selected with varying degrees of complexity and accuracy and will be described as follows:

The *Curtice Quadratic* model [3] was one of the first introduced for the simulation of nonlinear microwave FET devices. The drain current is described by (2.1) which contains four dc parameters to be determined, α, β, λ and V_P. The first defines the onset of drain-saturation voltage, and the second is proportional to the device transconductance. The third parameter takes account of the device's dc output conductance. Finally, the pinch-off voltage is defined by V_P.

$$I_{DS} = \beta(V_{GS} - V_P)^2 (1 + \lambda V_{DS})\tanh(\alpha V_{DS}) \quad (2.1)$$

An example of the transfer characteristic for this type of model is illustrated in Figure 2.6, adequate to analyze power amplifiers, mixers, and frequency multipliers. However, it is incapable of predicting intermodulation distortion, IMD, since there is no third order term in the equation.

The nonlinear capacitance within the active region is simulated by an ideal reverse-biased Schottky-junction diode, which is given by (2.2). The capacitance in the pinched off region is given by (2.3) which represents a constant capacitance. Such relations are expressed in graphical form in Figure 2.7.

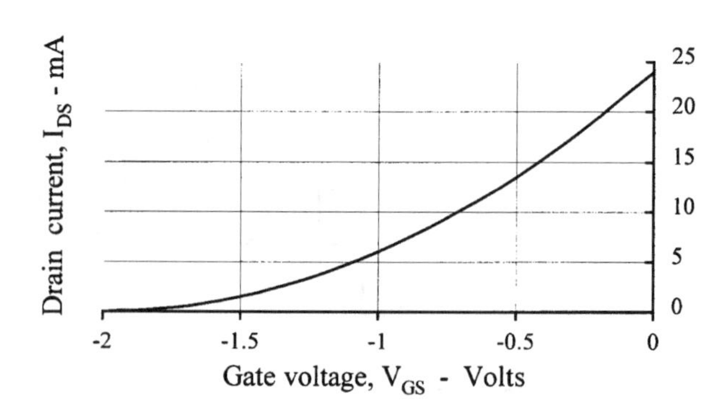

Figure 2.6 Transfer characteristic—Curtice Quadratic model.

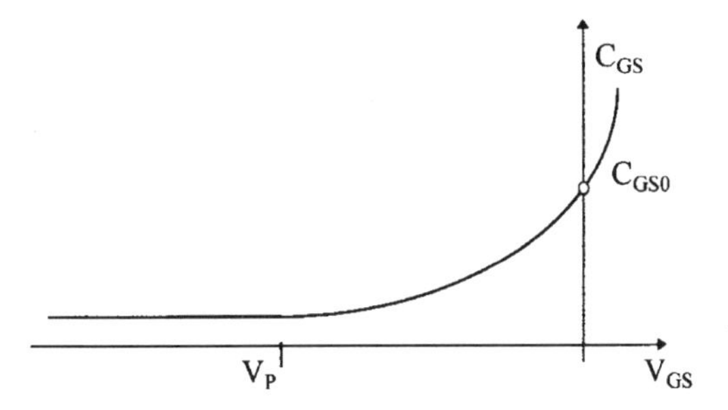

Figure 2.7 Relation between capacitance and voltage.

$$C_{gs} = C_{GS0}/\sqrt{1 - V_{GS}/V_{\phi}} \quad \text{for } V_P < V_{GS} \tag{2.2}$$

$$C_{gs} = C_{GS0}/\sqrt{1 + V_P/V_{\phi}} \quad \text{for } V_P > V_{GS} \tag{2.3}$$

where,

C_{GS0} = capacitance when $V_{GS} = 0$

V_{ϕ} = built in potential of the diode

The model assumes both C_{GS} and C_{GD} are given by the same equation, but with different constants and submitted to different reverse bias applied to each diode. The FET gate current I_G has two components: gate-source component, I_{GS}, and gate-drain component, I_{GD}, defined by (2.5) and (2.6), respectively. They also have different constants and are submitted to different reverse voltages during normal operation. At large signals, the gate-source diode might be forward biased in the positive cycle of applied signal and eventually might get into the verge of gate breakdown voltage, BV_{GS}, on the negative cycle. The gate-drain is always reverse biased in normal operation, it does not conduct in the forward direction but may conduct in the reverse direction, if the drain-source-applied voltage is greater than the breakdown voltage, BV_{DS}. The following set of equations assume that both diodes present similar breakdown characteristics.

$$I_G = I_{GS} - I_{GD} \tag{2.4}$$

$$I_{GS} = I_s[\exp(qV_{GS}/n_i kT_a) - 1] - I_{VB}[\exp(V_{GS} - BV_{GS})/V_0] \tag{2.5}$$

$$I_{GD} = I_s[\exp(qV_{GD}/n_i kT_a) - 1] - I_{VB}[\exp(V_{GD} - BV_{GD})/V_0] \tag{2.6}$$

where,

$$BV_{DS} = BV_{GS} - V_{DS} = BV_{GD} - V_{DS}$$

V_{GS} is the voltage across the gate-source diode
V_{GD} is the voltage across the gate-drain diode
I_s is the reverse saturation current of the Schottky barrier
I_{VB} is the leakage current due to breakdown voltage of the Schottky-barrier
V_0 is a voltage that describes the breakdown effect
n_i is the ideality factor

The *Curtice Cubic* model [4] is considered to be an improvement from the previous model where a cubic relation is introduced to take account of intermodulation in microwave circuits. The drain current is described by (2.7a) and (2.7b) where the following parameters define the current voltage relation, namely A_0, A_1, A_2, A_3, β_c, and α.

$$I_{DS} = (A_0 + A_1 V_1 + A_2 V_1^2 + A_3 V_1^3) \tanh(\alpha V_{DS}) \tag{2.7a}$$

$$V_1 = V_{GS} e^{-j\omega\tau} [1 + \beta_c (V_{DS0} - V_{DS})] \tag{2.7b}$$

where,

V_{DS0} = drain voltage at which the A_0, A_1, A_2, A_3 coefficients were determined

β_c = pinch-off voltage coefficient

τ = represents the time delay in the gate-source voltage

One of the problems of this model is on the parameters determination employing curve fitting procedures. The cubic relation may provide parameters resulting in a pinch-off that makes current zero, or transconductance zero, but not both simultaneously. It does not represent the real device. Observe that this equation also takes into account the change in the pinch-off voltage due to changes in the drain voltage. The simulated transfer characteristic for this model is based on the values of Table 2.1 obtained from a low current 1 μm gate device as depicted in Figure 2.8. The capacitances and the Schottky-barrier diodes in this model are simulated in a manner similar to the one employed in the quadratic model.

The originators of the *Statz-Pucel* model [5] recognized that a real MESFET device displays a transfer $I_{DS} - V_{GS}$ characteristic that resembles a linear device near zero gate volts and a quadratic one near the pinch-off voltage. This can be observed on the $I_{DS} - V_{GS}$ plot for the reference device, represented in Figure 2.9. With simple corrections to the previous equations for junction

Table 2.1
Elements and Parameter Values—Curtice Cubic MESFET Model

dc parameter	Value	Element	Value
A_0	0.016542 A	C_{gs}	0.527 pF
A_1	0.0500214 V^{-1}	C_{ds}	0.2513 pF
A_2	0.02012 V^{-2}	C_{gd}	0.087 pF
A_3	1.0×10^{-12} V^{-3}	R_g	2.9 Ω
α	2.16505 V^{-1}	R_s	2.4 Ω
β_c	–0.0394 V^{-1}	R_d	5.3 Ω
		R_{ds}	618.5 Ω
		BV_{DS}	15 V
		V_{DS0}	3.0 V
		I_s	1.0×10^{-9}A
		R_i	0.0 Ω

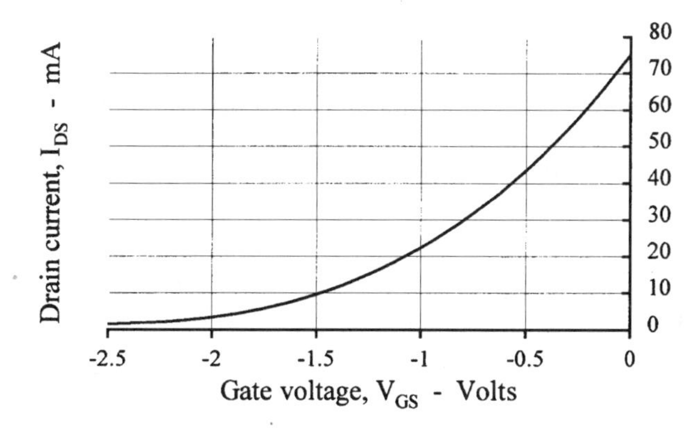

Figure 2.8 Transfer characteristic—Curtice Cubic model.

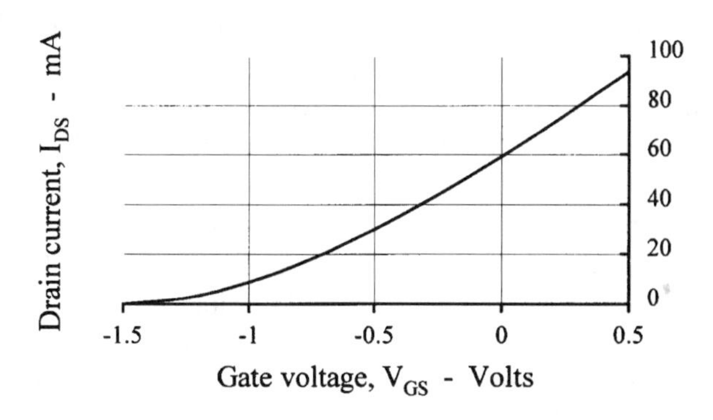

Figure 2.9 Transfer characteristic—Statz-Pucel model.

FETs, they derived a simple model that has received a wide acceptance among the design engineers.

The analytic description of this model is summarized in (2.8) as a product of two functions, one dependent entirely on gate voltage and the other on drain voltage. The first function describes the current source relation to the gate-source voltage assuming the device is operating under saturation. For gate voltages near pinch-off, the denominator of this function is close to unity and the function is quadratic. When the gate voltage is near zero volts, the denominator becomes a function of gate voltage which cancels the numerator quadratic law, resulting in a linear function for $F(V_{GS})$.

$$I_{DS} = F(V_{GS})G(V_{DS}) \tag{2.8}$$

$$F(V_{GS}) = \frac{\beta(V_{GS} - V_P)^2}{1 + B(V_{GS} - V_P)} \quad \text{for } V_P < V_{GS} < 0.7 \tag{2.9}$$

$$F(V_{GS}) = 0 \quad \text{for } V_{GS} < V_P \tag{2.10}$$

$$F(V_{GS}) = I_F \quad \text{for } V_{GS} > 0.7 \tag{2.11}$$

In the previous model, the drain current was considered an ideal source independent of drain voltage. However, real devices present a dependence which is described by function $G(V_{DS})$, represented by

$$G(V_{DS}) = \left[1 - \left(1 - \frac{\alpha V_{DS}}{3}\right)^3\right](1 + \lambda V_{DS}) \quad \text{for } 0 < V_{DS} < 3/\alpha \tag{2.12a}$$

$$G(V_{DS}) = 1 + \lambda V_{DS} \quad \text{for } V_{DS} \geq 3/\alpha \tag{2.12b}$$

The simulated output characteristics for the device described in Appendix A is illustrated in Figure 2.10. The linear region of the device is comprised between zero drain voltage and the knee voltage V_k ($\approx$ 1 volt). In this region, the device operates as a linear resistance whose value is controlled by the gate voltage. The knee voltage is defined by the coefficient α in (2.12).

Once the device is saturated, the current becomes a function of the drain voltage due to the output conductance of the device, given by the parameter λ. Observe that only five dc parameters, α, β, λ, B and V_P, are required by the model. The parameters for this model were determined for the reference device described in Appendix A, and are related in Table 2.2.

The authors also worked on defining capacitances capable of simulating the devices when operating in the normal mode when $V_{DS} > 0$, and on the reverse mode when $V_{DS} < 0$. In order to obtain the capacitance equations, they studied how the gate charge behaves during the voltage transitions introducing

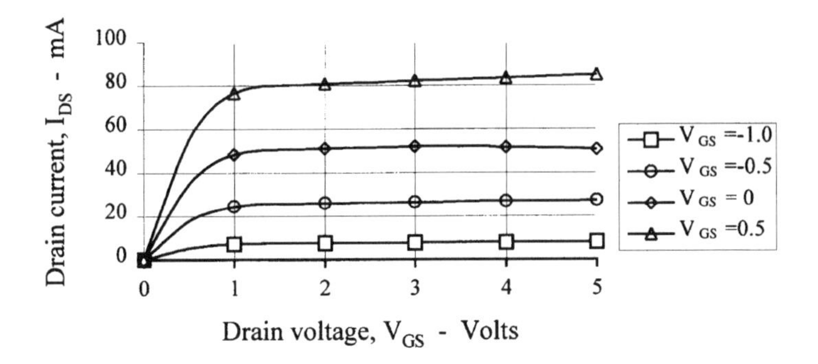

Figure 2.10 $I_{DS} - V_{DS}$ output characteristics—Statz-Pucel model.

Table 2.2
Elements and Parameter Values—Statz-Pucel MESFET Model

dc parameter	Value	Element	Value
α	2.035 V^{-1}	C_{gs}	0.50 pF
β	0.044 mA/V^2	C_{ds}	0.15 pF
B	0.29968 V^{-1}	C_{gd}	0.054 pF
λ	0.015 V^{-1}	R_g	4.0 Ω
γ	2.16505	R_s	3.0 Ω
V_P	−1.5 V	R_d	3.0 Ω
		R_{ds}	800.0 Ω
		BV_{DS}	7.0 V
		I_s	1.0×10^{-14}A
		R_i	0.0 Ω

parameters that avoided singularities and allowed smooth transitions. Differentiating the charge on the gate with respect to gate-source and gate-drain voltage, the following set of equations were obtained:

$$C_{gs} = \frac{0.25 C_{GS0}}{(1 - V_{NEW}/V_\phi)^{0.5}} f(\delta, V_P)\, f(\alpha, V_{GS}, V_{GD}) + 0.5 C_{GD0}\, f_o(\alpha, V_{GS}, V_{GD}) \tag{2.13}$$

$$C_{gd} = \frac{0.25 C_{GS0}}{(1 - V_{NEW}/V_\phi)^{0.5}} f(\delta, V_P)\, f_o(\alpha, V_{GS}, V_{GD}) + 0.5 C_{GD0}\, f(\alpha, V_{GS}, V_{GD}) \tag{2.14}$$

where,

$$V_{NEW} = 0.5^*\{V_{EFF} + V_P + [(V_{EFF} - V_P)^2 + \delta^2]^{0.5}\}$$

$$V_{EFF} = 0.5^*\{V_{GS} + V_{GD} + [(V_{GS} - V_{GD})^2 + (1/\alpha)^2]^{0.5}\}$$

$$f(\delta, V_P) = \left\{1 - \frac{V_{EFF} - V_P}{[(V_{EFF} - V_P)^2 + \delta^2]^{0.5}}\right\}$$

$$f(\alpha, V_{GS}, V_{GD}) = \left\{1 + \frac{V_{GS} - V_{GD}}{[(V_{GS} - V_{GD})^2 + (1/\alpha)^2]^{0.5}}\right\}$$

$$f_o(\alpha, V_{GS}, V_{GD}) = \left\{1 - \frac{V_{GS} - V_{GD}}{[(V_{GS} - V_{GD})^2 + (1/\alpha)^2]^{0.5}}\right\}$$

C_{GS0} = gate-to-source capacitance @ $V_{GS} = 0$
C_{GD0} = gate-to-drain capacitance @ $V_{DS} = 0$

The parameter δ was introduced to allow for a smooth transition in C_{GS} while entering into pinch-off. In reality it represents the range of voltages during which the gate voltage changes from V_{NEW} to V_P. Parameter α is the same one used to define the dc current. A qualitative plot of C_{GS} as a function of drain- and gate-voltage is depicted in Figure 2.11(a, b), for a gate measuring 1×100 μm. The model shows an excellent prediction of the MESFET capacitances in the normal and reverse mode of operation. However, if the device operates below pinch-off, C_{GS} reduces to zero introducing convergence problems when solving the circuit nonlinear equations.

In real devices, the capacitance in this region is small, but it is not zero. The reason for the nonconvergence can simply be explained by the inductance matching the gate capacitance. It is intuitive that the inductance will be delivering current to a nonexistent capacitance which is a particular circuit condition that has to be avoided. To overcome this problem, a constant low capacitance corresponding to the capacitance of the pinched-off channel, is added in parallel with C_{GS}.

Another problem [6] in this model is its inability to correctly represent the transistor output resistance, R_{DS}, the inverse of G_{DS}. In order to verify this statement let us obtain the output conductance in the saturation region by applying the definition

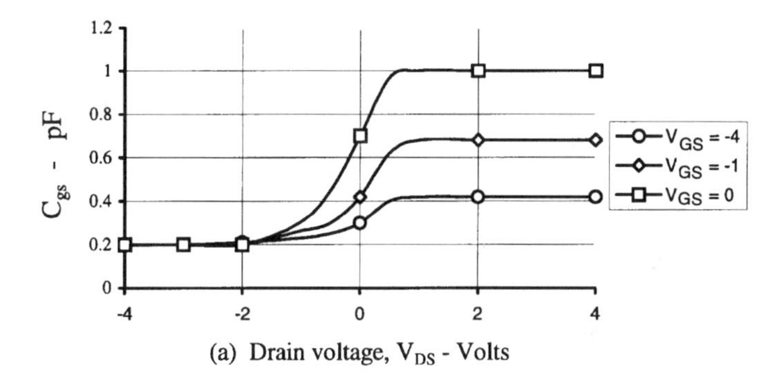

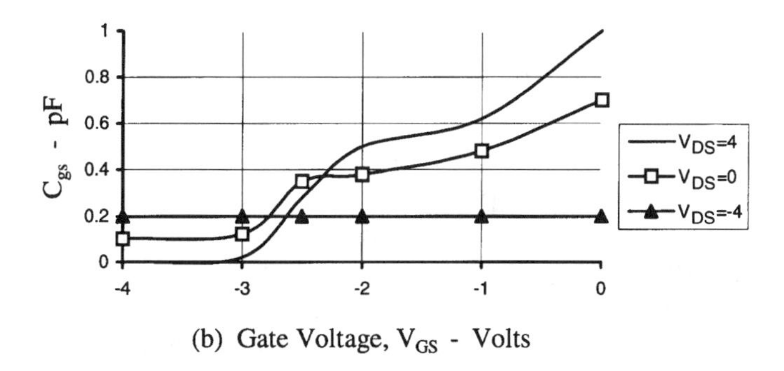

Figure 2.11 Gate-to-source capacitance; (a) in function of V_{DS}, and (b) in function of V_{GS}.

$$G_{DS}(V_{GS}, V_{DS}) = \frac{\partial I_{DS}}{\partial V_{DS}} = \frac{\beta(V_{GS} - V_P)^2\lambda}{1 + B(_{GS} - V_P)} \quad \text{for } V_{GS} > V_P \quad (2.15a)$$

$$G_{DS}(V_{GS}, V_{DS}) = 0 \quad \text{for } V_{GS} < V_P \quad (2.15b)$$

This equation shows that the parameter λ is chosen to fit the transistor $I_{DS} - V_{DS}$ curves at a specific gate voltage point $V_{GS} = V_{GS0}$. As V_{GS} increases above V_{GS0}, the model's R_{DS} becomes smaller than the real R_{DS}, and the reverse happens when the gate voltage is below V_{GS0}. Poor prediction of output conductance will affect the output impedance, thus S_{22}, and the device output power prediction.

The *Triquint Own model* (TOM) [7] was developed to improve the Statz-Pucel model and correct several of the described model deficiencies. The first correction is on the pinch-off voltage which is made a function of the drain voltage, according to (2.16) improving the output conductance at low drain currents. The relation between gate voltage and drain current is made with the arbitrary power law, q in Equation (2.17), giving more versatility to the model.

$$V_P = V_{P0} + \gamma_T V_{DS} \tag{2.16}$$

The second improvement is on the smoothing of the $I_{DS} - V_{DS}$ curve slope. Therefore, the drain current is defined by the set of equations described next, where the parameter δ_T allows the slope of the I–V curve to vary as a function of the drain current. It takes into account the power dissipated (junction temperature) by the device. In spite of the more accurate representation of the drain current, the model loses the simplicity of the original model, since three more parameters need to be fitted during the determination.

$$I_{DS0} = \beta(V_{GS} - V_P)^q \tag{2.17}$$

$$I_{DS} = \frac{I_{DS0}\tanh(\alpha V_{DS})}{(1 + \delta_T V_{DS} I_{DS0})} \tag{2.18}$$

The *Materka-Kacprzak* model [8] was developed to give a good simulation of the breakdown effects in power GaAs MESFET devices. Therefore, it is capable of giving a more accurate representation of saturation of the output power in a MESFET amplifier. The nonlinear elements from this model are the gate-to-source capacitance C_{GS}, the diodes D_{GD} and D_{GS} and the drain-current source, I_{DS}. The first three elements are defined as in the Curtice model. The drain current is described by (2.19) where the parameters I_{DSS}, α_M, V_{po} and γ_T define the current voltage relation. The voltage V_P is defined by (2.16).

$$I_{DS} = I_{DSS}\left(1 - \frac{V_{GS}}{V_P}\right)^2 \tanh\left[\frac{\alpha_M V_{DS}}{V_{GS} - V_P}\right] \tag{2.19}$$

A model example for the device 2SK273 [8], by Mitsubishi, whose parameters are described in Table 2.3, is represented in the $I_{DS} - V_{DS}$ plot shown in Figure 2.12.

2.4 Table-based Models

A table-based model [9–11] is not analytical since it does not use parametrized equations to represent the drain and gate current as a function of terminal voltages. Instead, a series of dc and S-parameter measurements are performed over a large combination of gate and drain bias to build up a table model. A mathematical treatment on this data generates a model capable representing dc, small-signal, and large-signal operation.

Table 2.3
Elements and Parameters for the Materka Model

dc parameter	Value	Element	Value
α_M	3.35	C_{gs}	0.64 pF
γ_T	$-0.11\ V^{-1}$	C_{ds}	0.1 pF
I_{DSS}	75.0 mA	C_{gd}	0.026 pF
V_P	-1.78 V	R_g	4.5 Ω
n_i	1.086	R_s	4.6 Ω
BV_{DS}	15.0 V	R_d	4.7 Ω
I_s	1.05×10^{-9} A	R_{ds}	∞ Ω
I_{VB}	6.5×10^{-9} A	R_i	10.0 Ω

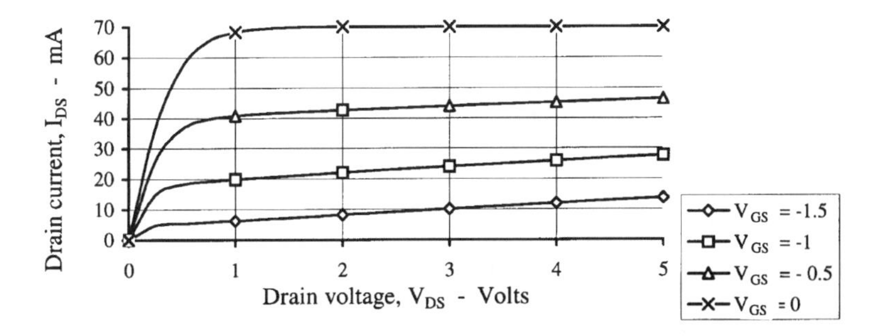

Figure 2.12 $I_{DS} - V_{DS}$ output characteristics—Materka model.

The Hewlett-Packard *Root* model [10], represented by a "black-box" in Figure 2.13, fits into this category. It contains eight functions (five of them are linearly independent), and each is a nonlinear function of the intrinsic terminal voltages.

The author proposes the use of 1000 measurement points, a value which is practical from a data acquisition and data storage viewpoint. The equations describing the drain and gate current are given by

$$I_{DS} = hI_{DS}^{DC}(V_{GS}, V_{DS}) + \frac{d}{dt}Q_D(V_{GS}, V_{DS}) + (1 - h)I_{DS}^{high}(V_{GS}, V_{DS}) \tag{2.20}$$

$$I_{GS} = I_{GS}^{DC}(V_{GS}, V_{DS}) + \frac{d}{dt}Q_G(V_{GS}, V_{DS}) \tag{2.21}$$

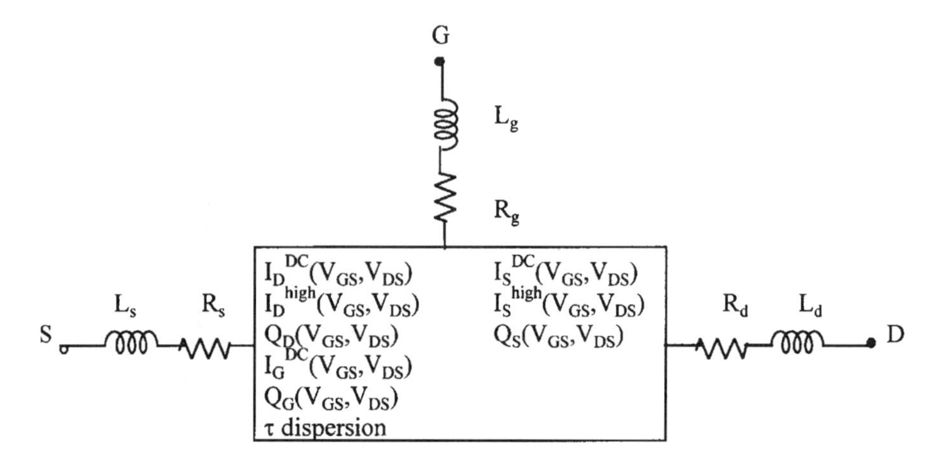

Figure 2.13 Root model.

where, h = is a function that characterizes the dispersion between dc and high frequency currents. It changes smoothly from unity at dc and zero at high frequencies.

$$Q_G(V_{GS}, V_{DS}) = f[I_{DS}^{DC}\text{meas}(V_{GS}, V_{DS}), S^{\text{meas}}(V_{GS}, V_{DS}, \omega), \omega, \ldots]$$

It can be observed that drain current is obtained not only from dc measurements but also from high-frequency measurements. Therefore, this is a *Non-Quasi-Static* model which means its elements are frequency-dependent. For instance, the transconductance will change from low to high frequency expressing a measured effect. Thus, the model accounts for trap effects that usually have a great impact on the use of dc parameters at high frequencies. The Root model offers the highest accuracy, since no fitting steps are involved and the results are based on real measurements. However, the model is less portable since it does not conform to simple analytical expression that can easily be incorporated in nonlinear simulators. It also requires expensive equipment for its characterization.

The accuracy, properties, dc equations, and performance for the models presented in this context are summarized [12] in Table 2.4. All of them have their own advantages and disadvantages. The Statz-Pucel model meets required goals of simplicity, reasonable accuracy and availability in most non-linear simulators, including Pspice® [13]. It was decided to employ this model in all subsequent simulations presented in this book.

Table 2.4
Summary of Model's Performance

Model	$I_{DS} = f(V_{DS}, V_{GS})$	**Comments**
Curtice Quadratic	$I_{DS} = \beta(V_{GS} - V_P)^2 (1 + \lambda V_{DS}) \tanh(\alpha V_{DS})$	Quadratic function precludes IMD simulation and mixer spurious response. Junction diode to represent C_{GS}, C_{GD}.
Curtice Cubic	$I_{DS} = (A_0 + A_1 V_1 + A_2 V_1^2 + A_3 V_1^3) \tanh(\alpha V_{DS})$, $V_1 = V_{GS} e^{-j\omega\tau} [1 + \beta_c (V_{DS0} - V_{DS})]$	Cubic function allows IMD simulation and spurious products. Takes account of pinch-off changes with drain voltage. C_{GS} and C_{GD} similar to quadratic. Better $G_{DS}(V_{DS})$ match than Statz.
Statz-Pucel	$I_{DS} = \dfrac{\beta(V_{GS} - V_P)^2}{1 + B(V_{GS} - V_P)} (1 + \lambda V_{DS}) [1 - (1 - \alpha V_{DS}/3)^3]$ when $V_{DS} > 3/\alpha$, the last term is replaced by 1.	$I_{DS} - V_{GS}$ quadratic near pinch-off, linear near I_{DSS}. Improved capacitance model, dependence on V_{DS}. Allows IMD simulation.
Materka-Kacprzak	$I_{DS} = I_{DSS}\left(1 - \dfrac{V_{GS}}{V_P}\right)^2 \tanh\left[\dfrac{\alpha_M V_{DS}}{V_{GS} - V_P}\right]$	Models change in pinch-off voltage with V_{DS}. Better $G_{DS}(V_{DS})$ than Statz.
TOM (Triquint's Own model)	$I_{DS} = I_{DS0}/(1 + \delta_T V_{DS} I_{DS0}) [1 - (1 - \alpha V_{DS}/3)^3]$, $I_{DS0} = \beta(V_{GS} - V_P)^q$, $V_P = V_{PO} + \gamma_T V_{DS}$ for $V_{DS} > 3/\alpha$ $I_{DS} = I_{DS0}/(1 + \delta_T V_{DS} I_{DS0})$	Modified Statz model with arbitrary power law for $I_{DS} - V_{GS}$. Improved $G_{DS}(V_{DS})$ fit. Uses Statz capacitances.
HP Root model	Table-based; behavior interpolated from data taken at many discrete points.	Circumvents curve-fitting problems. Highest accuracy.

2.5 Model Accuracy

2.5.1 Long Time Constants

The initial work in nonlinear modeling focused on the dc I–V curves of a FET, based on the assumption that the large signal device properties are governed primarily by the transistor dc characteristics. However, changes in the occupancy state at the interface substrate/active layer and also on the surface traps, cause the dc and RF I–V characteristics of GaAs MESFETs to be different. Both of them are important to simulate the operation of real circuits. At high frequencies, the long-time constant electron traps present in the device do not participate in the conduction process. This "trap state" is a function of the average drain voltage and current, (i.e., the average component of the voltage and current waveform at frequencies below a few MHz). Thus, after fitting the dc curves, one finds that the small-signal output conductance provided by the model is lower by a factor of two or more compared to RF measured values. The same problem is found in the small-signal transconductance. A rather simplistic correction to this situation is to place a linear resistor in series with a capacitor and in parallel with the device output impedance. It is a good approach as long as the output voltage swing is confined to the linear region. A more accurate approach replaces the fixed resistor by a nonlinear resistor function of terminal voltages.

2.5.2 Thermal Effects

A second issue in the validity of the dc I–V curves for nonlinear modeling is thermal effects. The analytic equation assumes an isothermal operation of the device. However, when collecting dc data, the channel temperature changes as a function of applied bias, resulting in different channel temperatures for each bias. Considering that the electron's mobility behaves in a manner inversely proportional to temperature changes, and that it has a direct impact on the drain current and errors in the modeling, especially for large gate devices, are probable. The coefficient of the hyperbolic tangent described in (2.18) can be used to correct the thermal effects on the I_{DS} current, by adjusting the parameter δ_T.

2.5.3 Pulsed Measurements

The previous effects plus short-channel and Gunn-domain effects, which are not accurately predicted even by the most sophisticated physics-based models, have lead to the use of pulsed measurements by some researchers. Comparison of the I–V characteristics acquired under static and pulsed conditions seems to correct the thermal effects and bypass the trap effects in the device. The negative output conductance measured at dc in power devices, for instance, becomes positive when measured under pulsed conditions. This method of operation also allows determination of a dc transconductance and output conductance that matches the RF performance measured in the same manner. Therefore, this method is also capable of offsetting the long time constants associated with the I–V determination. In spite of the complexity of such measurement systems, they are preferred for characterization of high power devices.

2.6 Determination of the Model

The conventional method of extracting the equivalent circuit element values for a given device involves a set of dc and S-parameter measurements. The dc steps are employed to determine the nonlinear drain-source current generator as a function of external bias, the drain-source and drain-gate properties, as well as to extract the device parasitic resistances. This set of data is used to adjust the model parameters, so that it can simulate, within a certain degree of accuracy, the same measured dc currents as a function of applied voltages. Optimization techniques are usually employed to fit theoretical response to the measured ones. The model can be any one of the popular nonlinear models: Statz-Pucel, Curtice, Materka, TOM, etc. The S-parameters are employed to determine the device capacitances, wire bond inductances, and to correct the differences between dc and RF nonlinear current behavior. Recently, new procedures [14] have been introduced, where no dc measurements are performed to extract the device parasitics. They consist of S-parameter measurements with a "cold device," (i.e., with an unbiased device). However, the conventional technique presented below is simpler, requires the use of a curve tracer, a vector network analyzer, and curve-fitting software routines. These techniques provide complete characterization of all elements, providing reasonable accuracy for the design of frequency multipliers and determination of $I_{DS} - V_{DS}$, $I_{DS} - V_{GS}$ and $I_{GS} - V_{GS}$ plots.

2.6.1 DC Measurements

Schottky-barrier Diodes

The MESFET structure of Figure 2.1 contains two Schottky-barrier diodes, one connecting the gate-to-source and the other connecting the gate-to-drain.

The gate-source diode has its current described by (2.5) and the gate-drain diode by (2.6).

Gate-source Diodes

The diode parameters, I_s and n_i, can be extracted by plotting gate current versus gate voltage with the drain open circuited. The gate current versus voltage for an open-drain FET model is equal to

$$V_{GS} = (R_S + R_G)I_{GS} + \frac{n_i k T_a}{q} \ln\left[\frac{I_{GS}}{I_s}\right] \tag{2.22}$$

This equation can be simplified if a low current is injected into the diodes so that the series resistances can be neglected. Then, using (2.23), one can transform the exponential $I_{GS} - V_{GS}$ plot into a linear plot as depicted in Figure 2.14. This result was taken from a device whose gate dimensions are 0.5×250 μm.

$$\ln(I_{GS}) = \ln(I_s) + \frac{qV_{GS}}{n_i k T_a} \tag{2.23}$$

In this figure, notice that at high voltages the current deviates from the linear condition due to the series resistances ($R_G + R_S$) that can no longer be neglected. At low voltages the current deviates from the diode law due to leakage and effects of current recombination in the semiconductor.

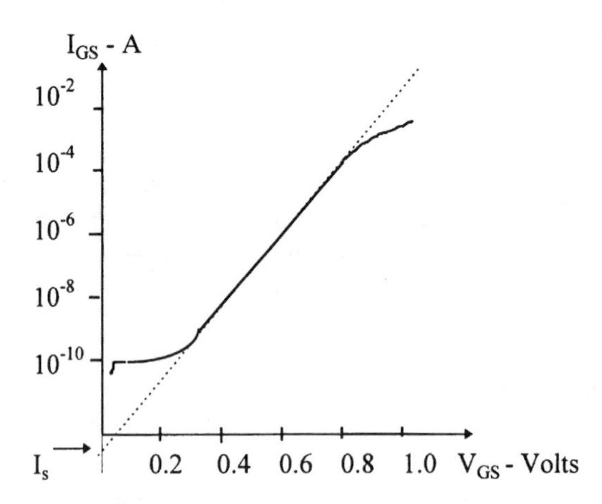

Figure 2.14 Gate-source diode: $I_{GS} - V_{GS}$.

The slope from the linear part of this plot provides information on the ideality factor described by (2.24). One can also obtain the diode saturation current by extrapolating the linear portion to $V_{GS} = 0$.

$$n_i = \frac{\partial V_{GS}}{\partial[\ln(I_{GS})]} \tag{2.24}$$

Gate-drain Diodes

A rough estimation [15] of the avalanche current can be obtained from the device's I–V plot at high drain voltages and low gate voltage, as shown in Figure 2.15. The difference between the extrapolated drain current from low-drain voltage to high voltage and the measured current gives an estimation of avalanche current in the device.

A better accuracy is obtained by pulsing the gate and drain voltage and measuring the gate current. An example of such a plot for a $0.5 \times 800\mu\text{m}$ device is shown in Figure 2.16.

2.6.2 Parasitic Series Resistances

One of the techniques [16, 17] used to determine the series resistances is to apply a constant current to the gate and measure the corresponding gate voltage for three conditions: drain open, source open, and drain and gate shorted. Considering the static model for a FET illustrated in Figure 2.17, the gate current is given by (2.22). Observe that in this condition $I_{DS} = 0$.

If the gate current is high, then the second term of the equation becomes negligible and only the first term is considered. According to Figure 2.14, high injection starts at a few mA for this device. Thus, applying a current from 5 to 10 mA, one obtains the following equations:

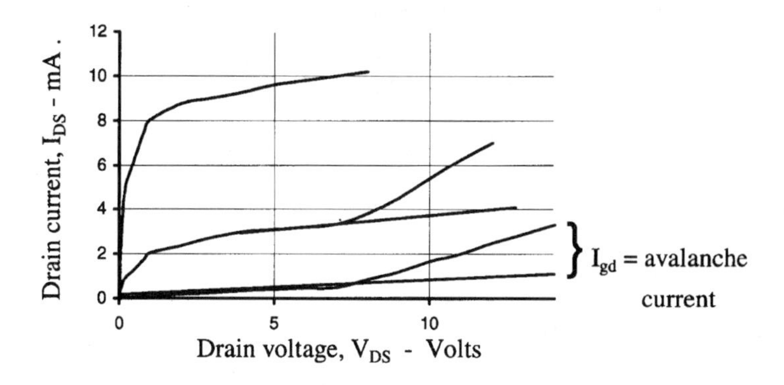

Figure 2.15 Measured FET static I–V curves.

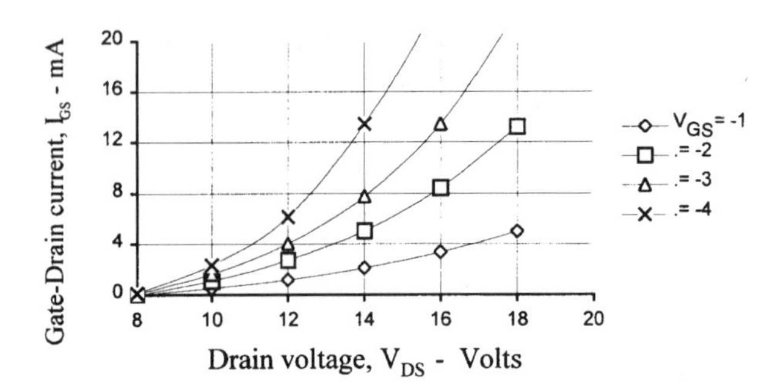

Figure 2.16 Gate-drain current versus drain voltage.

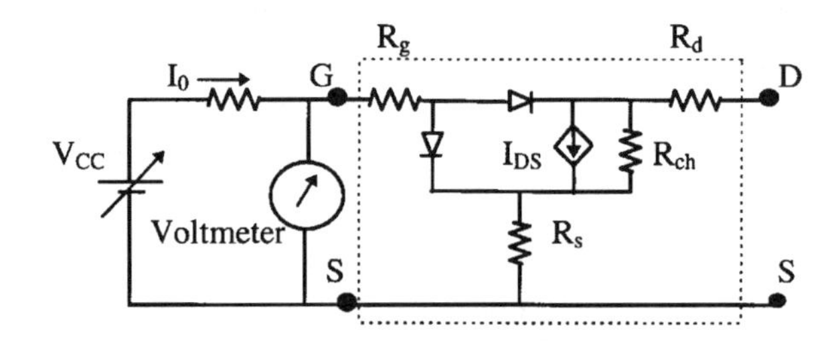

Figure 2.17 Determination of series resistances.

$$V_{GS} = (R_s + R_g)I_{GS} \qquad \text{(open drain)} \tag{2.25}$$

$$V_{GD} = (R_d + R_g)I_{GD} \qquad \text{(open source)} \tag{2.26}$$

$$V_{GDS} = \left[R_g + \frac{R_s R_d}{R_s + R_d} I_{GDS} \right] \qquad \text{(drain and source shorted)} \tag{2.27}$$

These equations can be solved for R_s, R_d, and R_g, by means of (2.28) and (2.29), where $I_{GS} = I_{GD} = I_{GDS} = I_0$.

$$R_s = \Delta R_1 + \sqrt{\Delta R_1^2 + \Delta R_1 \Delta R_2} \tag{2.28}$$

$$R_d = \Delta R_2 + R_s \tag{2.29}$$

where,

$$\Delta R_1 = (V_{GS} - V_{GDS})/I_0$$

$$\Delta R_2 = (V_{GD} - V_{GS})/I_0$$

Once the drain and source resistors are determined, the gate resistance can be determined either from (2.25) or (2.26).

An alternative method to determine the parasitic resistance, R_s, consists of an injection of a large current on the gate-to-source diode, as shown in Figure 2.18, so that the second term of (2.22) can again be neglected. The measured voltage on the floating drain V_o, gives the voltage on the source resistor, V_i, and consequently, $R_s = V_o/I_o$. The drain resistor is determined in a similar manner by grounding the drain.

In Figure 2.1, both the described methods have neglected the flow of current through the channel. Therefore, the measured source or drain resistors include half the value of the channel resistance, R_{ch}. However, at high gate current, its value is much smaller than either R_s or R_d, and can be neglected. The gate metalization behaves like a resistive transmission line at high frequencies so that the measured dc resistance has a different effective value at high frequencies. An empirical procedure is to divide the dc result by three in order to get the RF value, (2.30). A more accurate value is found by curve fitting to high frequency measurements.

$$R_g = (R_{meas} - R_s)/3 \tag{2.30}$$

2.6.3 Drain Current Function of Terminal Voltages

In this step, static measurements are made on a curve tracer to determine the transfer characteristics $I_{DS} - V_{GS}$ at a given drain bias, and the output characteristics $I_{DS} - V_{DS}$ as a function of gate voltage. The experimental plots have to be modified to account for the effect of series resistance, R_s and R_d. Their effect is not important at low currents but cannot be neglected at high drain currents. The device static model of Figure 2.19 can be used to derive the set of equations that take these resistances into account.

The external static currents are dependent on the current through the diodes and the drain-source current source, and are defined by (2.31) to (2.33).

$$I_G = I_{GD}(V_{GDi}) + I_{GS}(V_{GSi}) \tag{2.31}$$

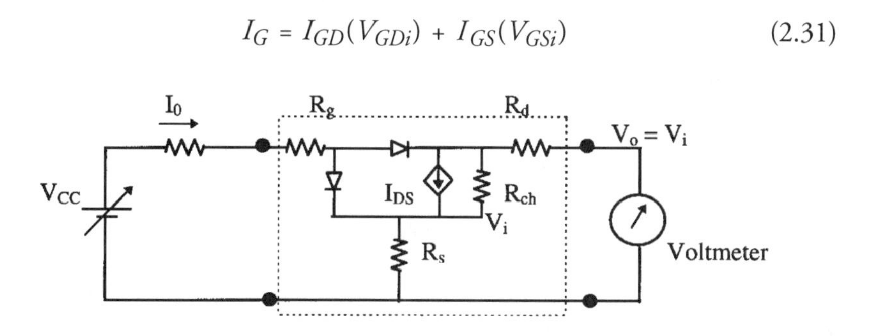

Figure 2.18 Alternative determination of series resistance.

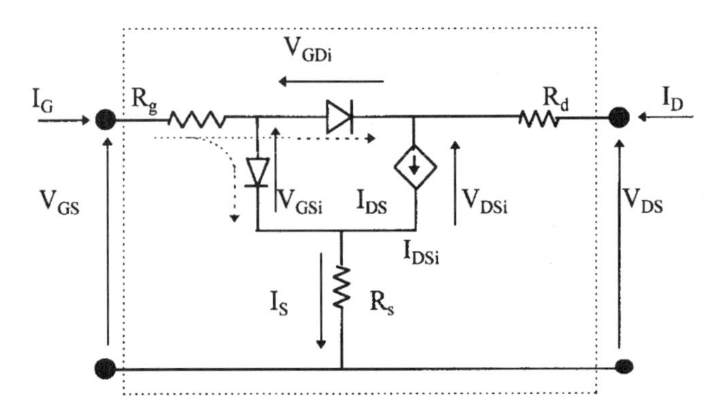

Figure 2.19 Static dc model.

$$I_D = I_{DSi}(V_{GSi}, V_{DSi}) - I_{GD}(V_{GDi}) \quad (2.32)$$

$$I_S = I_{DSi}(V_{GSi}, V_{DSi}) - I_{GS}(V_{GSi}) \quad (2.33)$$

The relation between the external and internal voltages is given by the set (2.34) to (2.36).

$$V_{GSi} = V_{GS} - R_g I_G - R_s I_S \quad (2.34)$$

$$V_{GDi} = V_{GD} - R_g I_G - R_d I_D \quad (2.35)$$

$$V_{DSi} = V_{DS} - R_d I_D - R_s I_S \quad (2.36)$$

The model parameters are determined by using the resistors R_s, R_d, R_g, the $I_{DS} - V_{GS}$ and $I_{DS} - V_{DS}$ plots and curve fitting techniques to make any one of the analytic models simulate the measured dc characteristics.

2.6.4 RF Measurements

The purpose of RF measurements is to determine the reactances and parasitics of the model. They also are useful in the correction of frequency effects on drain current, transconductance and output conductance.

The first step consists of the determination of the nonlinear reactances by measuring S-parameters as a function of terminal voltages [11]. For this purpose, the S-parameter measurements are carried out at low frequencies, (i.e., in the range 0.1 to 1 GHz). At these low frequencies, the wire bonding inductances and the transit time can be neglected, and the measurements are free from frequency dispersion effects due to surface states. Therefore, the mathematical description of the model elements as a function of S-parameters can be greatly simplified. The optimum measurement frequency range is determined by checking the band in which the magnitude of S_{21} and S_{22} remains constant, an indication of validity of the model.

Initial measurements are carried out for the simplified model represented in Figure 2.20. For this model it is possible to demonstrate [11] that the relation between the model parameters and S-parameters are given by (2.37) to (2.41) described next:

$$g_{ds}' = \left|\frac{1 - |S_{22}|}{1 + |S_{22}|}\right|^{\gamma_s} \tag{2.37}$$

$$g_m' = 0.5|S_{21}|(1 + g_{ds}') \tag{2.38}$$

$$C_{gd}' = \frac{1}{2\omega}|S_{12}|(1 + g_{ds}') \tag{2.39}$$

$$C_{gs}' = -\left\{\frac{1}{2\omega}\phi(S_{11}) + \frac{C_{gd}'(1 + g_{ds}' + g_m')}{1 + g_{ds}'}\right\} \tag{2.40}$$

$$C_{ds}' = -\left\{\frac{1}{2\omega}\phi(S_{22})\left[(1 - g_{ds}')^2\right] + C_{gd}'(1 + g_m')\right\} \tag{2.41}$$

where,

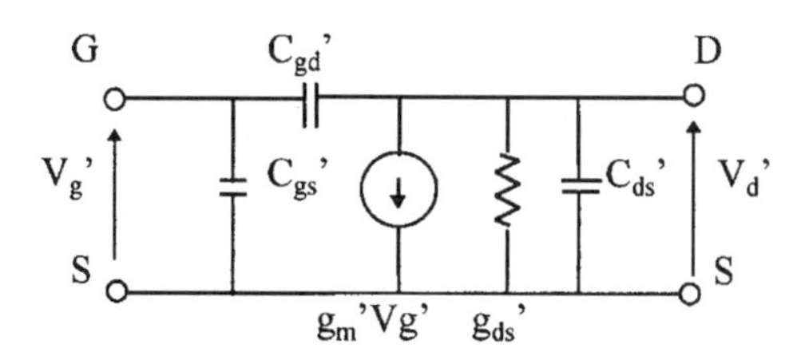

Figure 2.20 Simplified model for extraction of elements.

$\phi(S_{11})$ and $\phi(S_{22})$ are respectively the phase of parameters S_{11} and S_{22}, in radians

$\gamma_s = -1$ if $0° < \phi(S_{22}) < 180°$

$\gamma_s = 1$ if $-180° < \phi(S_{22}) < 0°$

$g_{ds}' = g_{ds}^*/Y_0$, $g_m' = g_m^*/Y_0$, $C_{gd}^* = C_{gd}/Y_0$

$C_{gs}' = C_{gs}^*/Y_0$, $C_{ds}' = C_{ds}^*/Y_0$

Y_0 = reference characteristic admittance

The coefficient γ_s in (2.37) determines if the drain impedance is a conductance in parallel with a capacitance, case of $\gamma_s = 1$, or inductance in series with a resistance, when $\gamma_s = -1$. The particular case of $\phi(S_{22}) = 0$ or 180 degrees results in a pure resistance or conductance. The previous normalized parameters are obtained directly from the device terminals, and have to be corrected to include the effect of the device series resistance, depicted in Figure 2.21, by employing (2.42) to (2.47).

$$D_0^* = R_s(g_m^* + g_{ds}^*) + R_d g_{ds}^* \tag{2.42}$$

$$g_m = g_m^*(1 - D_0^*) \tag{2.43}$$

$$g_{ds} = g_{ds}^*/(1 - D_0^*) \tag{2.44}$$

$$C_{gs} = C_{gs}^* + (g_m^* + g_{ds}^*)(R_s C_{gs}^* - R_d C_{gd}^*)/(1 - D_0^*) \tag{2.45}$$

$$C_{gd} = C_{gd}^* - g_{ds}^*(R_s C_{gs}^* - R_d C_{gd}^*)/(1 - D_0^*) \tag{2.46}$$

$$C_{ds} = C_{ds}^* - R_g[C_{gd}^*(g_m^* + g_{ds}^*) + C_{gs}^* g_{ds}^*] \tag{2.47}$$

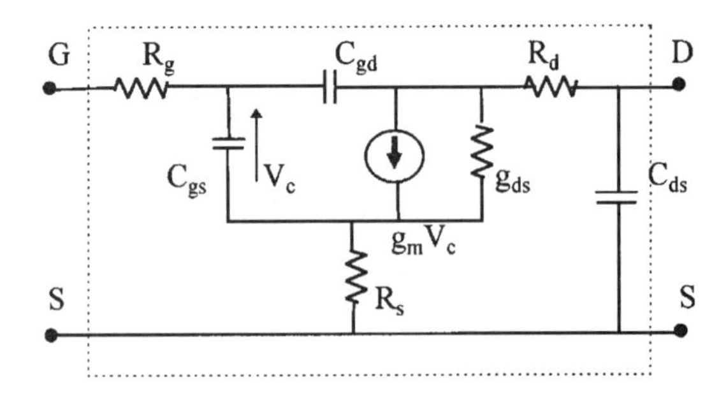

Figure 2.21 Simplified model including series resistances.

2.6.5 S-parameter Measurements as a Function of Bias

This procedure is applied for a large combination of bias voltages in order to map the behavior of reactances on the input and output plane. However, this tedious job can be simplified, making only two bias voltage combinations represented in Figure 2.22, one with constant V_{DS} and the other with constant V_{GS}. This simplification provides the necessary information for the analysis of most nonlinear circuits. After this step, the small-signal parameters from the model are available as a function of external bias.

The resulting RF transconductance and output conductance obtained are used to correct the dc value, thus keeping consistency of the model over frequency. For small-signal operation, a simple series RC circuit can be added in parallel with the current source. For large signal simulations, that approach is no longer valid and a more complex solution has to be employed. In this case, a simple way to take account of the dispersion effect is to include in the model an extra current source [18] in parallel with I_{DS}, as depicted in Figure 2.23. The reactance of capacitor C_{RF} determines the frequency where the high frequency correction takes place.

The second step in the characterization process is determining the remaining parasitic elements, such as wire bond inductances, charging resistor for capacitance C_{gs}, R_i, and delay time associated with transconductance, τ. This step can be performed at a single bias in a first approximation of the final model, but the S-parameters have to be taken from very low frequencies to the highest frequency of interest. Now, the full model is considered, all parasitics are added to the model of Figure 2.24, and curve-fitting techniques are applied only to these elements so that the best fit of the modeled S-parameters matches the measured one with reasonable accuracy.

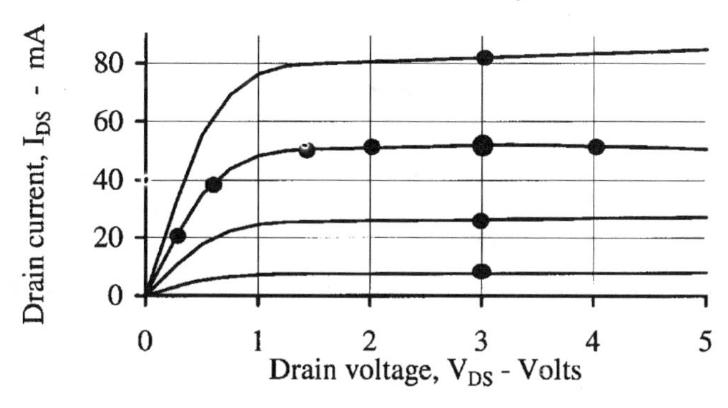

Figure 2.22 Selection of bias voltages for S-parameter measurements.

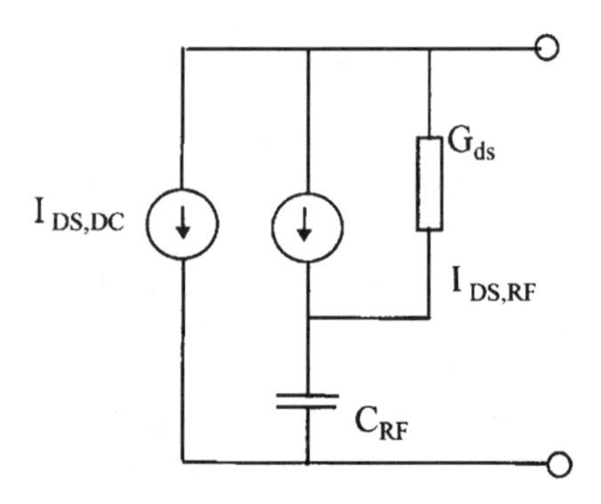

Figure 2.23 Correction of dispersion on g_m and g_{ds}.

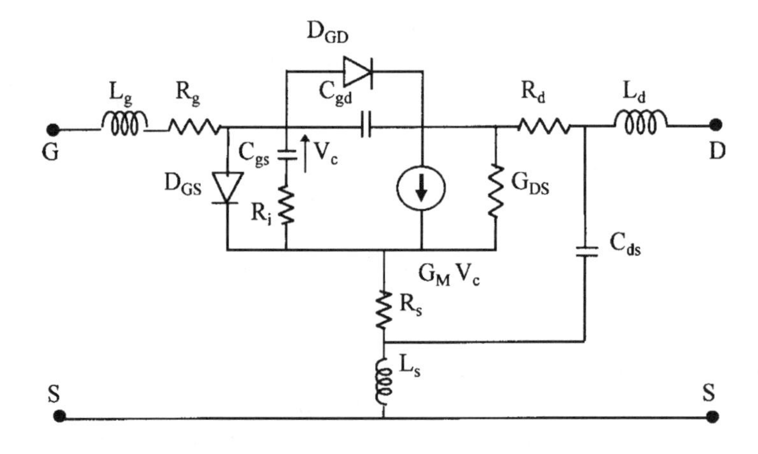

Figure 2.24 Total nonlinear equivalent model.

References

[1] Tajima, Y., B. Wrona, and K. Mishima, "GaAs FET Large-Signal Model and its Application to Circuit Designs," *IEEE Transactions on Electron Devices,* Vol ED 28, No. 2, February 1981, pp 171–175.

[2] Smith, H., and A. W. Swanson, "HEMTs—Low Noise and Power Transistors for 1 to 100 GHz," *Applied Microwaves,* May 1989, pp. 63–72.

[3] Curtice, W., "A MESFET Model for Use in the Design of GaAs Integrated Circuits," *IEEE Transactions on Microwave Theory and Techniques,* Vol MTT-28, No. 5, May 1980, pp 448–456.

[4] Curtice, W., and M. Ettenberg, "A Nonlinear GaAs FET Model for Use in the Design of Output Circuits for Power Amplifiers," *IEEE Transactions on Microwave Theory and Techniques,* Vol MTT-33, No. 12, December 1985, pp 1383–1394.

[5] Statz, H., P. Newman, I. M. Smith, R. A. Pucel, and H. A. Haus, "GaAs FET device and circuit simulation in SPICE," *IEEE Transactions on Electron Devices,* Vol ED-34, Feb. 1987, pp 160–169.

[6] Correra, F. S., "Development of Nonlinear Models for PHEMTs," Private report.

[7] Triquint Scaleable Nonlinear GaAs FET EETOM1, Series IV HP-EESOF, *Microwave & RF Circuit Design Manual: Circuit Element Catalog—Nonlinear Devices,* Vol. 1, Chapter 12.

[8] Materka, A. and T. Kacprzak, "Computer Calculation of Large-Signal GaAs FET Amplifier Characteristics," *IEEE Transactions on Microwave Theory and Techniques,* Vol MTT-33 No. 2 , February 1985, pp. 129–135.

[9] Willing, H. A., C. Rauscher, and P. de Santis, "A Technique for Predicting Large-Signal Performance of a GaAs MESFET," IEEE Trans, *On Microwave Theory and Techniques,* Vol MTT-26 No. 12 , December 1978, pp. 1017–1023.

[10] Root, D., S. Fan, and J. Meyer, "Technology Independent Large Signal Non Quasi-Static FET Models by Direct Construction from Automatically Characterized Device Data," 21st European Microwave Conference, Stuttgart, Germany 1991, pp. 927–930.

[11] Perichon, R. A., P. Gouzien, and E. Camargo, "Determination experimentale d'un schema equivalent non lineaire simple representatif du TEC AsGa," *L'Onde Electrique,* Vol. 64, No. 2, Mars–Avril 1984, pp. 79–85.

[12] McKay, T.—Private Communication, 1993.

[13] *Pspice User's Manual Version 5.4,* MicroSim Corp., CA, July 1993.

[14] Vogel, B., "The Application of RF Wafer Probing to MESFET Modeling," *Microwave Journal,* November 1988, pp. 153–162.

[15] Hwang, V., and T. Itoh, "An Efficient Approach for Large-Signal Modeling and Analysis of the GaAs MESFET," *IEEE Transactions on Microwave Theory and Techniques,* Vol. MTT-35, No. 4, April 1987, pp. 396–402.

[16] Kurita, O. and K. Morita, "Microwave MESFET Mixer," *IEEE Transactions on Microwave Theory and Techniques,* Vol. MTT-24, No. 6, June 1976, pp 361–366.

[17] Fukui, H., "The Determination of the Basic Device Parameters of a GaAs MESFET," *The Bell System Technical Journal,* Vol. 58, No. 3, March 1979, pp. 771–797.

[18] Cojocaru, V., and T. J. Brazil, "Scaleability of dc/AC Nonlinear Dispersion Models for Microwave FETs," 1997 IEEE-MTT International Microwave Symposium, June 1997, Denver, CO, pp. 387–390.

3

Low Frequency Multipliers

Computer simulation of nonlinear circuits is an invaluable tool to study complex microwave topologies in a practical manner. However, its application requires the design engineer to know how the nonlinear elements behave under large-signal conditions. In order to give such an insight, this chapter employs simple low frequency models and graphical analysis to help visualize the fundamental concepts of frequency multipliers. The convenience of low-frequency models relies on the use of resistive nonlinearities which are relatively easier to understand. The nonlinear reactances are neglected here and are considered later in the chapter on high frequency circuits. From this low-frequency analysis, the best compromise is between harmonic generation, device bias, and drain termination for a given multiplication ratio.

The nonlinear model has been covered extensively in the previous chapter. The simplified low frequency FET model employed in this chapter is represented in Figure 3.1. It contains a nonlinear current source, the gate-source and gate-drain diodes.

Three sources of low frequency nonlinearity are shown in this model, but the current source is the most effective in the generation of harmonics. The Schottky-barrier diodes could be used for the same purpose, but they introduce excessive loss in this process during conduction. They also compromise reliability. The current source may generate even harmonics if an asymmetrical distortion is induced in the output current or voltage waveform. Odd harmonics are generated by inducing, at the output, distortions on both peaks of either waveform. Those effects are controlled by bias and proper harmonic drain termination.

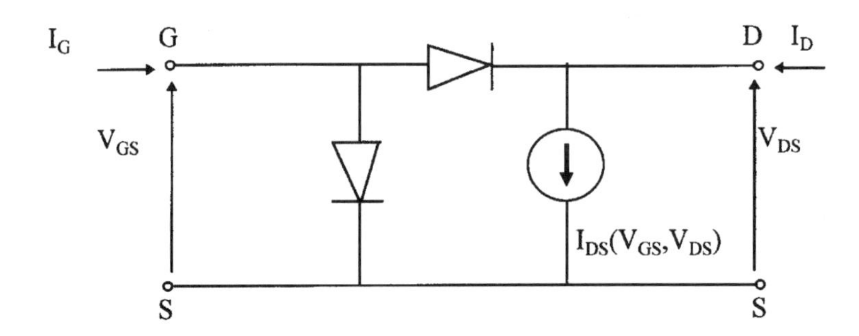

Figure 3.1 Low frequency model of a FET.

In order to gain insight into the mechanism of harmonic generation, one can apply simple analytical models to the current source and study the harmonics generated by a high-dynamic voltage applied to the gate.

3.1 Application of a Piecewise Linear Model

In this section, the current source of the model represented at the input plane in Figure 3.1 is used to describe the drain current as a function of gate voltage according to (3.1). That relation is assumed to be linear for gate voltages within the interval from pinch-off voltage to zero volts. A graphical representation is given by the plots of Figure 3.2. The diodes in the model are assumed to be ideal, being a short circuit for voltages greater than zero and an open circuit for voltages lower than zero. For gate voltages lower than pinch-off, there is no drain current. For voltages greater than zero, the drain current is assumed

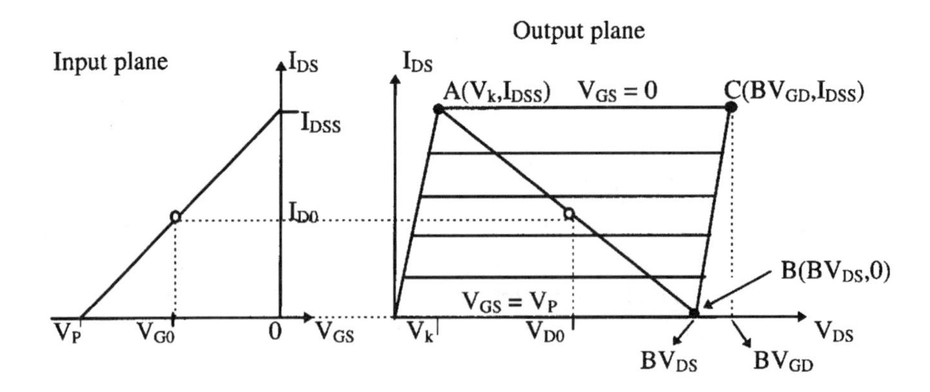

Figure 3.2 Piecewise linear model of a FET.

to be constant and equal to I_{DSS} due to the conduction of the gate-to-source diode.

$$I_{DS} = I_{DSS}\,(1 - V_{GS}/V_P) \quad \text{for } V_P < V_{GS} < 0 \tag{3.1a}$$

$$I_{DS} = 0 \quad \text{for } V_{GS} < V_P \tag{3.1b}$$

$$I_{DS} = I_{DSS} \quad \text{for } V_{GS} > 0 \tag{3.1c}$$

The important points to consider in the output plane are:

Point A. The drain current is equal to I_{DSS} and the low drain voltage is in the limit between active and resistive region, defined by the knee voltage, V_k.

Point B. There is no drain current because the device is pinched-off and the drain voltage is on the limit of breakdown of the gate-to-drain diode. The maximum drain voltage that can be applied is defined as BV_{DS}.

Point C. The gate voltage is zero so that the drain current is maximum, and the drain voltage is on the limit of breakdown of the gate-to-drain diode. The maximum drain voltage that can be applied is defined as BV_{GD}.

Observe that maximum drain voltage is dependent on the gate voltage. At point B, the drain voltage and gate voltage are both applied to the gate-to-drain diode, while at point C, only the drain voltage is applied. Therefore the drain voltage is more restrictive at low drain currents. The model also assumes that drain current is independent of drain voltage, which is valid as long as the drain voltage is kept within the limits $V_k < V_{DS} < BV_{DS}$.

The application of a periodic signal at the gate will induce at the output a current that will be dependent on the type of drive waveform, bias point and drain terminations. Such a large-signal effect can be studied by applying Fourier analysis to the resulting currents and voltages, which gives relationships to the various frequency components and their phases.

The generic circuit in Figure 3.3 can represent a low-frequency amplifier or a frequency multiplier. The circuit contains a matching resistor at the gate

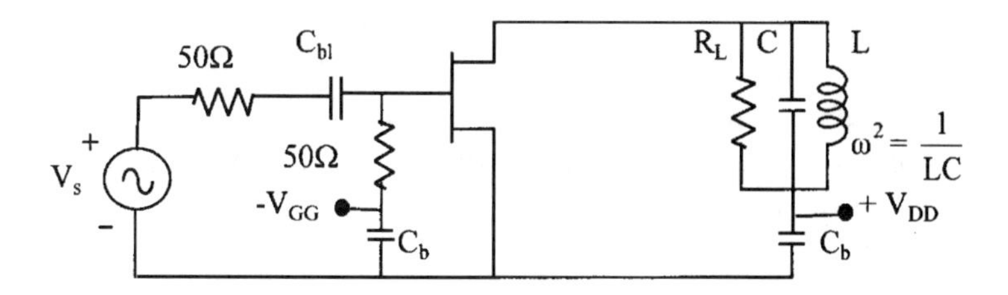

Figure 3.3 Generic low frequency amplifier/multiplier.

side, and a resonant LC filter connected to the drain side tuned to the output signal. The matching resistor makes the gate voltage equal to half the generator voltage in this circuit. The blocking capacitor, C_{bl}, blocks the dc current but is assumed to be a short circuit at the operating frequency. The capacitor C_b is also a short circuit at the operating frequencies and bypasses the power supply impedance as well as any noise that eventually couples to the bias line. When representing a tuned amplifier, the resonant circuit is tuned at the fundamental frequency. If a frequency multiplier is under consideration, then the tank circuit is tuned at the output harmonic.

3.1.1 Class A

A class A amplifier is defined as one where the quiescent gate voltage is adjusted so that the drain current is set to $I_{D0} = I_{DSS}/2$ and drain voltage is set to $V_{D0} = (BV_{DS} - V_k)/2$. Let us also assume a resistive termination crossing the points $A(I_{DSS}, V_k)$ and $B(0, BV_{DS})$. This is the bias and load for normal linear operation of FET devices. Assuming the gate voltage comprises a dc bias voltage, V_{G0}, and a sinusoidal voltage of amplitude V_g, ($V_g < V_P/2$), there is no generation of harmonics in the drain current described by (3.3), due to the linear relation between drain current and gate voltage. The maximum linear operation for the gate voltage and drain current is illustrated in Figure 3.4, where the time origin is made coincident with the maximum signal amplitude.

$$V_{gs}(t) = V_{G0} + V_g \cos \omega t \tag{3.2}$$

$$I_{ds}(t) = I_{D0} + I_{d1} \cos \omega t \tag{3.3}$$

where,

I_{d0} = dc component of drain current

I_{d1} = fundamental frequency component of drain current

In the limit where the peak of the drain current becomes equal to $I_{DSS}/2$, the waveform is still sinusoidal. To find the magnitude of the gate voltage and the time or angle, $\varphi_0 = \omega_0 t$, for the drain current to reach I_{DSS}, starting from the quiescent point, one can inspect the plots, or obtain that value applying the proper conditions to (3.1).

$$I_{ds}(\varphi_0) = I_{DSS}\left[1 - \frac{V_{G0} + V_g \sin(\varphi_0)}{|V_P|}\right] \tag{3.4}$$

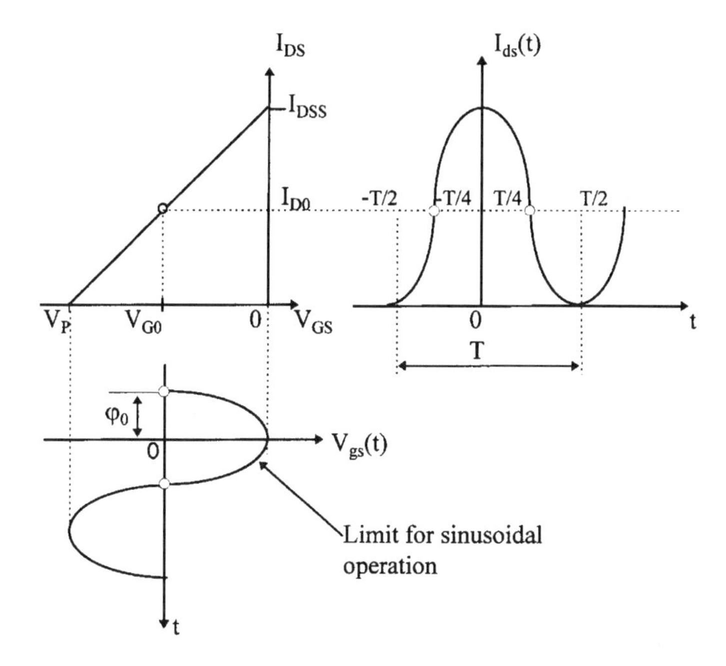

Figure 3.4 Linear class A operation.

Solving (3.4) for the angle φ_0 and assuming that $V_{G0} = |V_P|/2$ and $I_{ds}(\varphi_0) = I_{DSS}$, one obtains,

$$\varphi_0 = \sin^{-1}(2V_g/|V_P|) \tag{3.5}$$

Therefore, if $V_g = |V_P|/2$, then, $\sin\varphi_0 = 1$, and the angle required for the drain current to reach I_{DSS} is equal to $\pi/2$, corresponding to the waveforms in Figure 3.4.

Increasing the gate-peak voltage beyond the linear limits, the drain current waveform will change, as observed in Figure 3.5. The angle φ_0 will become lower than $\pi/2$. Increasing the gate peak amplitude, the angle φ_0 will become lower and lower. In the limit, $\sin(\varphi) = \varphi$, where the angle $\varphi = \omega t$. Then the waveform of Figure 3.5 is trapezoidal. The time origin is again selected at the maximum amplitude. The selection of a different origin will result in the same magnitude for the generated harmonics, but their phases will be different [2]. In this context the time origin is selected such that it corresponds to maximum amplitude, generating a cosines Fourier series. Then, (3.3) takes the form,

$$I_{ds}(t) = I_{D0} + I_{d1}\cos(\omega)t + I_{d2}\cos(2\omega)t + \ldots = \sum_{n=0}^{\infty} I_{dn}\cos(n\omega t) \tag{3.6}$$

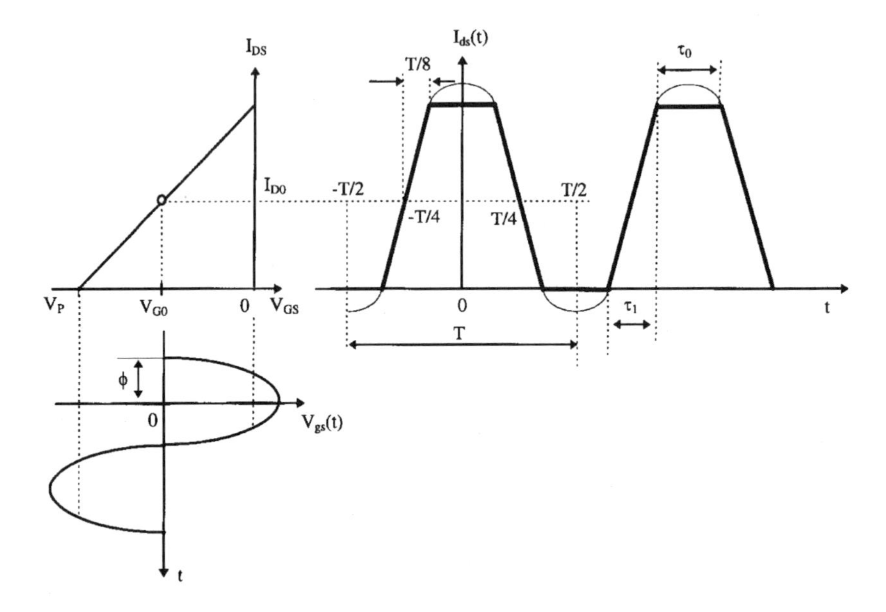

Figure 3.5 Trapezoidal wave shape generation.

The particular case of a symmetrical trapezoid is obtained when $\tau_0 = \tau_1 = T/4$. The gate voltage that enables the drain current to be a symmetrical trapezoid can be calculated, observing that the time or phase of the sinusoidal portion of the distorted drain current from $-T/4$ up to the trapezoid horizontal line is equal to $t_1 = T/8$, or

$$\varphi_1 = \omega t_1 = \frac{2\pi}{T}\frac{T}{8} = \frac{\pi}{4} \tag{3.7}$$

The desired gate voltage is found by inserting (3.7) into (3.5) and solving for V_g. The resulting value, as expected, is greater than half the pinch-off voltage.

$$V_g = \frac{|V_P|}{\sin(\pi/4)} = 0.707|V_P| \tag{3.8}$$

Looking into the waveform, it is intuitive that the dc term does not change, and the harmonic content will depend on how the sinusoidal wave will be clipped. In the case of symmetrical trapezoids, the Fourier coefficients are tabulated [1] and are given by (3.9a) and (3.9b).

$$I_{D0} = I_P\left[\frac{\tau_0 + \tau_1}{T}\right] \tag{3.9a}$$

$$I_{dn} = 2I_p\left[\frac{\tau_0 + \tau_1}{T}\right]\frac{\sin(n\pi\tau_1/T)}{n\pi\tau_1/T}\frac{\sin[n\pi(\tau_0 + \tau_1)/T]}{n\pi(\tau_0 + \tau_1)/T} \tag{3.9b}$$

where,

$(\tau_0 + \tau_1)$ = conduction time

T = period of $I_{ds}(t)$

I_{dn} = nth harmonic of drain current

I_p = peak amplitude of drain current

The ratio $(\tau_0 + \tau_1)/T$ is also known as the duty cycle, and is equal to 0.5 for a symmetrical trapezoidal waveform. In (3.9) all even harmonics result in zero amplitude, thus, this wave shape contains only odd harmonics. As the amplitude of the applied voltage increases further, τ_1 becomes zero, and the output current waveform will approach that of a square wave. This is the condition where the amplitude of odd harmonics is maximum. Therefore, waveforms of this type are adequate to build odd harmonic multipliers, like triplers, quintuplers, etc.

Employing the circuit of Figure 3.3 as a saturated amplifier, the filter has the objective of rejecting all generated harmonics, and the amplifier delivers a pure sinusoidal at the output. Amplifiers in this mode of operation are used for amplification of FM modulated signals, or a certain type of phase-modulated signals where there is negligible amplitude modulation. If the circuit is employed as a frequency multiplier, then the filter has to reject all harmonics, except the output one.

An important parameter in FET circuits is the transconductance defined as the partial derivative of the drain current with respect to the gate voltage, around a dc bias point.

$$g_m = \frac{\partial I_{DS}}{\partial V_{GS}} = -\frac{I_{DSS}}{V_P} \tag{3.10}$$

A similar parameter is defined under large-signal operation, the large-signal transconductance, G_M, where it is defined by the ratio of ac drain current to ac gate voltage. Note that in this condition, the large and small-signal transconductance are identical.

$$G_M = I_{d1}/V_g = -I_{DSS}/V_P \tag{3.11}$$

Replacing the device in Figure 3.3 by the model from Figure 3.1, it can be shown that the voltage gain for class A operation, G_{VA}, is given by

$$G_{VA} = \frac{I_{d1}}{V_g} R_L = G_M R_L \tag{3.12}$$

The voltage gain is constant for any drive signal with amplitude $V_g < V_P/2$. If the input sinusoidal signal has an amplitude $V_g > V_P/2$, the output waveform will resemble a trapezoid and from this point on, the device becomes saturated. Observe that the magnitude of the drain current was increasing proportionally to the gate voltage before saturation. Now, the fundamental frequency component of the drain current, I_{d1}, is no longer proportional but rather given by the fundamental frequency content on the output waveform. Thus, the large-signal transconductance ($G_M = I_{d1} / V_g$) decreases when the device enters saturation, due to higher increase in gate voltage compared to the I_{d1} increase.

Driving the device with a square pulsed signal, the output will also be a pulsed signal which is rich in harmonics. But in this case there is no contribution from the device in the generation of harmonics and it will not be considered in this context. It is, however, important in the design of high-efficient power amplifiers. In such a case, the signal in the output plane is either on point A (V_k,I_{DSS}) or point B (BV_{DS},0) which are low power dissipation points. The transition from one point to the other crosses regions of higher dissipation but that happens only during a short time interval compared to the signal period. Therefore, the average power in one cycle of the signal is low, resulting in low dc power dissipated in the device, and an amplifier with a high-drain efficiency defined by $\eta = P_{out}/P_{DC}$, after filtering the harmonics.

3.1.2 Classes B and C

Harmonics can also be generated if other biasing schemes are employed. Before discussing class B and C operations, it is important to define the angle of conduction, (2ϕ), which corresponds to the angular portion of the sinusoidal gate cycle, during which the drain current is greater than zero. That conduction angle, expressed by (3.13) and with parameters shown in Figure 3.6, is directly dependent on device bias and signal amplitude. The higher the signal amplitude, the lower the conduction angle. The lower the value of V_{gmin}, the higher the conduction angle. Class A operation is defined when the conduction angle is equal to 360 degrees. True class B is defined when the device is biased at $V_{G0} = V_P$, and the conduction angle is equal to 180 degrees.

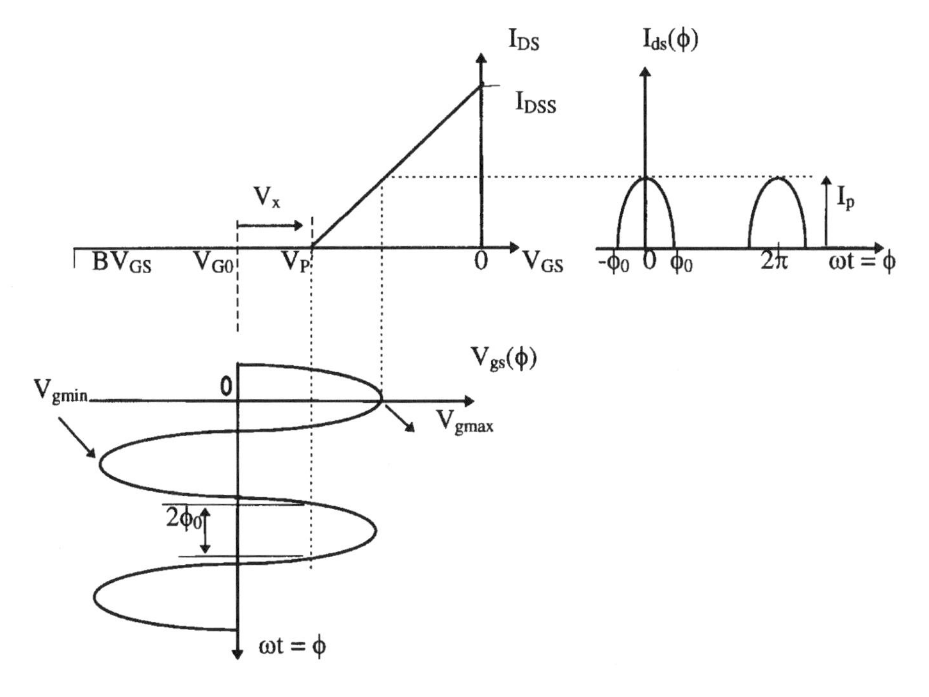

Figure 3.6 Definition of conduction angle.

$$\cos(\phi) = \frac{V_x}{V_g} = \frac{V_P - V_{G0}}{V_g} = \frac{2V_P - (V_{gmax} + V_{gmin})}{V_{gmax} - V_{gmin}} \tag{3.13}$$

where,

$$V_{G0} = (V_{gmax} + V_{gmin})/2 \tag{3.13a}$$

$$V_g = (V_{gmax} - V_{go}) = (V_{gmax} + V_{gmin})/2 \tag{3.13b}$$

Class AB is defined for a conduction angle between 180 and 360 degrees. Class C is defined for $V_{G0} < V_P$, where the angle of conduction is smaller than 180 degrees. Note that the gate voltage in this mode of operation can be as high as BV_{GD}, where the gate-source diode reaches breakdown. For class B operation, the maximum gate amplitude V_g is such that the total gate voltage is lower than BV_{GS} on the negative portion of the cycle and less than 0.0V on the positive side.

The safe region of operation in the drain side is determined by the type of load employed. The FET current source delivers current to the load, therefore, a low resistance load develops a low drain voltage, and a high resistance develops a high drain voltage. If the load is inductive, the output voltage may increase beyond the desirable maximum depending on the frequency of operation. In any case, the minimum drain voltage is V_k and the maximum is limited by the device drain-gate breakdown voltage, BV_{DS}. In low-frequency operation, the phase of drain and gate voltages are 180 degrees so the maximum voltages on the gate-side corresponds to minimum on the drain-side, and vice versa so that there is no maximum on both sides simultaneously. Thus, in low-frequency multipliers, it is possible to apply high dynamic voltages to the gate and drain and still maintain the device within a safe area of operation.

Applying a sinusoidal gate voltage, the drain current is composed of a train of linear sine-wave tips [2] whose Fourier terms are given by

$$I_{d0} = \frac{I_p}{\pi} \frac{\sin(\phi) - 2\phi\cos(\phi)}{(1 - \cos(\phi))} \tag{3.14}$$

$$I_{d1} = \frac{I_p}{\pi} \frac{\phi - \cos(\phi)\sin(\phi)}{(1 - \cos(\phi))} \tag{3.15}$$

$$I_{dn} = \frac{2I_p}{\pi} f(\phi)_n \tag{3.16}$$

where,

$$I_p = I_{DSS}(1 - V_{gs}/V_P)$$

$$f(\phi)_n = \frac{\sin(n\phi)\cos(\phi) - \mathrm{n}\sin(\phi)\cos(n\phi)}{n(n^2 - 1)(1 - \cos(\phi))}, \; n \geq 2$$

In the case of a class B amplifier, observe that the gate voltage has to be doubled in order to obtain the same drain current provided by class A, therefore, the class B voltage gain G_{VB}, is half the gain obtained from class A operation.

$$G_{VB} = \frac{I_{d1}R_L}{2V_g} = \frac{G_{VA}}{2} \tag{3.17}$$

The standard procedure to determine the best conduction angle for a given frequency multiplier is found by calculating the derivative of the specific harmonic current in respect to the conduction angle. In the case of frequency doublers, the optimum conduction angle is $2\phi = 120$ degrees. However, it becomes too complex to analytically determine the optimum angle for $n > 2$, so that it is easier to plot I_n/I_p as a function of conduction angle and graphically determine the best conditions. Such a plot is found in Figure 3.7 for conduction angles from 0 to 360 degrees.

It is interesting to observe that the optimum conduction angle is approximately given by the inverse of the multiplication ratio, $2\phi_{\text{opt}} = 240°/n$, and the harmonic current component by $I_{dn} = (0.54I_p)/n$ for $n > 1$. A summary of multiplication ratio and conduction angles normalized to the peak value of drain current is described in Table 3.1 for the first five harmonic components. For each angle, the table shows the value of each harmonic component. The table shows that for a conduction angle of 180 degrees, there is no odd harmonics, but for a square wave type of waveform, the table shows only odd harmonic components.

The small-signal transconductance is not defined for class B or C operation, but the large-signal value is still given by the ratio I_{d1}/V_g. That parameter

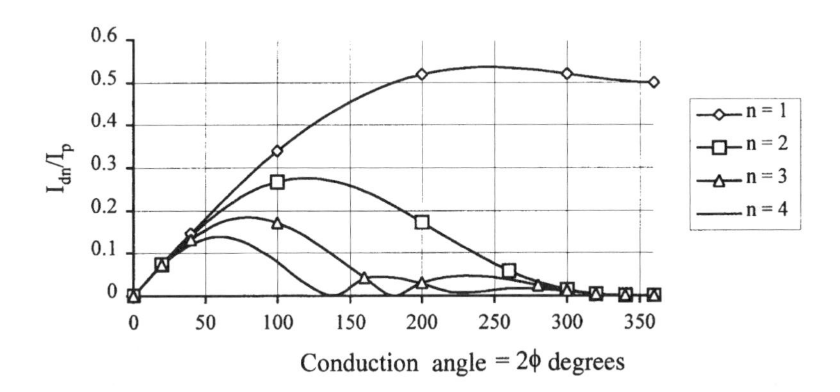

Figure 3.7 Normalized harmonic current as a function of conduction angle.

Table 3.1
Fourier Components as Function of Conduction Angle

		Fourier Frequency Components				
n	$2\phi_{opt}$	I_{d1}/I_p	I_{d2}/I_p	I_{d3}/I_p	I_{d4}/I_p	I_{d5}/I_p
1	360°	–6.0 dB				
2	120°	–8.4 dB	–12.0 dB	–17.0 dB	–32.0 dB	—
3	76°	–12.0 dB	–13.2 dB	–15.4 dB	–18.4 dB	–24.0 dB
4	65°	–13.4 dB	–15.4 dB	–15.9 dB	–18.0 dB	–20.0 dB
5	48°	–14.7 dB	–14.9 dB	–16.4 dB	–18.2 dB	–18.4 dB
pinch-off	180°	–6.0 dB	–13.2 dB	—	–27.0 dB	—
square wave	360°	–3.9 dB	—	–13.5 dB	—	–17.9 dB

can be employed in an equivalent large-signal circuit for the fundamental frequency and is useful for calculating large-signal impedance.

3.2 Application of a Square Law Model

In this model, the current source of Figure 3.1 assumes a drain current proportional to the square of the gate voltage within the range of gate voltages between pinch-off and zero volts. The drain current beyond those limits is defined as in the previous model. The I–V plots for this model are in Figure 3.8 and the analytic description is given by

$$I_{DS} = I_{DSS}(1 - V_{GS}/V_P)^2 \quad \text{for } V_P < V_{GS} < 0 \tag{3.18a}$$

$$I_{DS} = 0 \quad \text{for } V_{GS} < V_P \tag{3.18b}$$

$$I_{DS} = I_{DSS} \quad \text{for } V_{GS} > 0 \tag{3.18c}$$

The main difference in this model compared to the previous one is its capability of generating second harmonics in either class A, B, or C. The output current for a periodic sinusoidal drive will be again dependent on bias point and drain terminations.

3.2.1 Class A

The bias for class A operation is similar to the one described in the previous model, (i.e. defined for a drain current set equal to $I_{DSS}/2$, with $V_{G0} = V_P/3$). Some authors [4] prefer to define class A operation when $V_{G0} = V_P/2$ resulting in $I_{DSS}/4$ for the bias current. Assuming the gate voltage is described by (3.2) and is applied to the gate, the drain current is given by

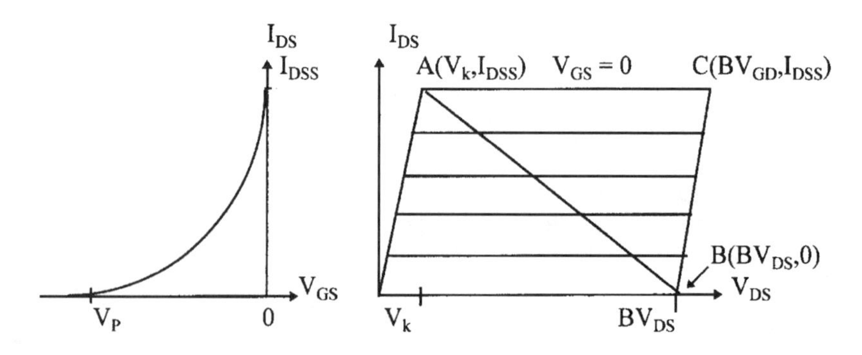

Figure 3.8 Square Law model of a FET.

$$I_{ds}(t) = I_{dn} = I_{D0} + I_{d1}\cos(\omega t) + I_{d2}\cos(2\omega t) \tag{3.19}$$

where,

$$I_{D0} = I_{DSS}\left\{(1 - V_{G0}/V_P)^2 + \frac{1}{2}(V_g/V_P)^2\right\} \quad \text{dc component}$$

$$I_{d1} = \frac{2I_{DSS}}{V_P}V_g(1 - V_{G0}/V_P) \quad \text{Fundamental frequency}$$

$$I_{d2} = \frac{I_{DSS}}{2}(V_g/V_P)^2 \quad \text{Second harmonic}$$

Note that the large-signal transconductance is again equal to the small-signal transconductance, which are defined by

$$g_m = \frac{\partial I_{DS}}{\partial V_{GS}} = -\frac{2I_{DSS}}{V_P}(1 - V_{G0}/V_P) \tag{3.20}$$

$$G_M = \frac{I_{d1}}{V_g} = -\frac{2I_{DSS}}{V_P}(1 - V_{G0}/V_P) \tag{3.21}$$

If the input voltage swings within the limits $V_P < V_{GS} < 0$, and a resistive load is applied to the drain, then a large second-harmonic-current component is present in the output waveform. Linear amplification then requires the use of a filter connected to the drain to eliminate the second harmonic voltage. In the case of a frequency doubler, a harmonic load is employed to reject fundamental frequency and matches the second harmonic to the output termination. Assuming that $V_{G0} = V_P/2$ and $V_g = V_P/2$, the ratio between the second harmonic current and the fundamental frequency is 1/4, which in decibels corresponds to 12 dB. The main advantage of a class A frequency doubler built with a square law device is a clean output spectrum with no generation of unwanted harmonics.

3.2.2 Classes B and C

A better second harmonic efficiency is obtained by biasing the device in class B. True class B is defined when the device is biased at $V_{G0} = V_P$, where the conduction angle is equal to 180 degrees. Now, the drain current is composed of a train of square law sine-wave tips where the first three Fourier terms are given by

$$I_{D0} = \frac{I_p}{\pi} \frac{\phi - 3\sin(2\phi)/4 + \phi\cos(2\phi)/2}{(1 - \cos(\phi))^2} \tag{3.22}$$

$$I_{d1} = \frac{2I_p}{\pi} \frac{3\sin(\phi)/4 + \sin(3\phi)/12 - \phi\cos(\phi)}{(1 - \cos(\phi))^2} \tag{3.23}$$

$$I_{d2} = \frac{2I_p}{\pi} \frac{\phi/4 - \sin(2\phi)/6 + \sin(4\phi)/48}{(1 - \cos(\phi))^2} \tag{3.24}$$

$$I_{dn} = \frac{2I_p}{\pi} \frac{(4 - n^2)\sin(n\phi) + (n-1)(n-2)\sin(n\phi)\cos(2\phi) + 3n\sin[(n-2)\phi]}{n(n^2 - 1)(n^2 - 4)(1 - \cos(\phi))^2} \tag{3.25}$$

where,

$$n \geq 3$$

$$I_p = \frac{I_{DSS}}{V_P^2}[V_g - (V_P - V_{G0})]^2$$

The maximum fundamental frequency current obviously is obtained for an angle of conduction equal to 360 degrees, when (3.23) reduces to a class A current. Maximum second harmonic current as a function of conduction angle is obtained by taking the derivative of (3.24) with respect to ϕ, which makes a conduction angle approximately equal to 135 degrees. The normalized Fourier coefficients as a function of conduction angle are shown in Figure 3.9,

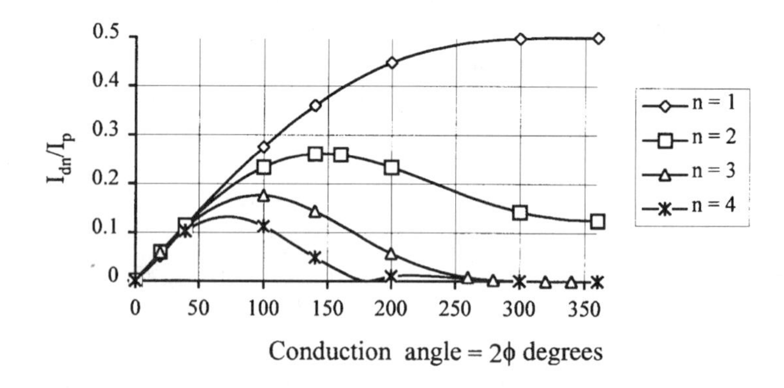

Figure 3.9 Normalized harmonic current as a function of conduction angle.

which are useful to determine the best conduction angle for a given multiplication ratio.

Two simplified MESFET models have been introduced. The first assumes a linear relation between drain current and gate voltage, and is a good approximation of real devices. The models of Chapter 2 show that the device is linear for most of the transfer characteristic, and is square law only in the vicinity of pinch-off. Thus the square law model is useful only if the device is biased near pinch-off and the applied signal amplitude is low. For this reason, an emphasis will be given to the linear model, which permits a simple evaluation of harmonic current and voltage in the circuit. In spite of the approximations, it provides good estimation of output power when operating as a power amplifier and harmonic power when operating as a frequency multiplier. More accurate results require the use of the models in Chapter 2.

3.3 Harmonic Power

In the previous sections it was demonstrated how to generate a drain current rich in harmonics from a sinusoidal voltage applied to the gate. In order to assess harmonic power, let us initially consider the thermodynamic balance of power in a FET [3], here considered as a dc to RF converter, represented by Figure 3.10.

In this nonlinear power converter, the power balance in the frequency domain is given by

$$P_{\text{in}} + P_{DC} = P_{\text{out}} + P_{\text{diss}} \tag{3.26}$$

In low-frequency circuits, there is very little power being absorbed by the gate of a FET, so that $P_{\text{in}} \approx 0$. Therefore, the dc power applied to the device is partially converted to RF energy at the output, and partially dissipated internally as heat.

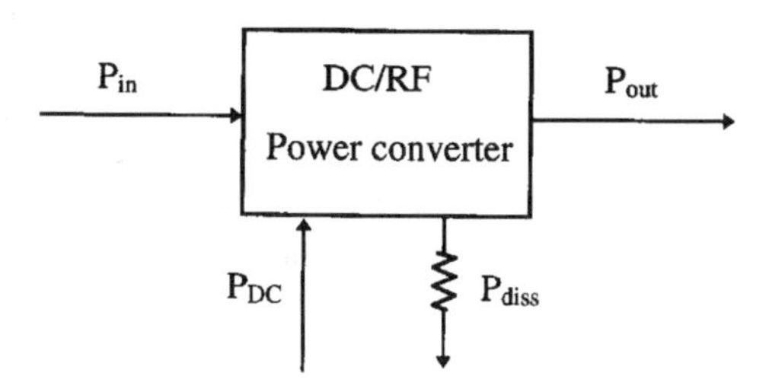

Figure 3.10 Thermodynamic frequency domain representation of an RF converter.

Dissipated power can be determined by considering the harmonic currents and voltages in the FET [4]. The harmonic currents can be expressed in the time domain by the sum of all frequency components, according to a Fourier series for a periodic signal, represented by

$$I_{ds}(t) = \sum_{n=-\infty}^{n=\infty} I_{dn} e^{jn\omega t} \tag{3.27}$$

Combining the drain current components with the load impedance obtains the drain voltage in the frequency domain through the equation:

$$\begin{aligned} V_{D0} &= V_{DD} \text{ at } n = 0 \\ V_{dn} &= -I_{dn} Z_L(n\omega), \ n = \text{output harmonic} \\ &\quad 0 \text{ for all other harmonics} \end{aligned} \tag{3.28}$$

This equation is valid for arbitrary loads, as long as the maximum voltage swing is within the range V_k and BV_{DS}, and as long as the drain current is independent of drain voltage. The drain voltage on the time domain is given by

$$V_{ds}(t) = \sum_{n=-\infty}^{n=\infty} V_{dn} e^{jn\omega t} \tag{3.29}$$

The power dissipated by the device in the time domain, P_{DISS}, within one cycle of the microwave signal is given by

$$P_{\mathrm{DISS}} = \frac{1}{T} \int_0^T I_{ds}(t) V_{ds}(t) dt \tag{3.30}$$

According to Parseval's Theorem [5], the dissipated power can also be obtained from the frequency domain components, as described by (3.31). This theorem implies that the total average power of a signal is composed of the sum of average power of each frequency component.

$$P_{\mathrm{DISS}} = \sum_{n=-\infty}^{n=\infty} I_{dn} V_{dn}^{*} \tag{3.31}$$

where, the value with an asterisk represents the complex conjugate of the original variable. The voltages and currents are root mean square values (rms) and should be transformed to peak values, V_{dn} and I_{dn}, a common practice in microwave circuits. An additional modification can be done in this equation, considering that $I_{ds}(t)$ and $V_{ds}(t)$ are real time functions. In this case, the harmonic currents and voltages are symmetrical conjugate in the frequency domain, so that (3.31) can be simplified to:

$$P_{\mathrm{DISS}} = \sum_{n=-\infty}^{n=\infty} P_{dn} \tag{3.32}$$

where,

$$P_{dn} = \begin{cases} I_{D0} V_{D0}, \; n = 0 \\ Re\{I_{dn} V_{dn}^{*}\}/2, \; n = \text{output harmonic} \\ 0 \text{ for all other harmonics} \end{cases}$$

In this equation, V_{D0} represents the dc supply voltage and I_{D0} is the time average value of the $I_{ds}(t)$ waveform, which is a function of bias, input signal drive, and waveform shape. The equation also states that, ideally there is no power dissipated in any harmonic except at the desired output frequency. If the device delivers power to the external circuit at the harmonic frequency instead of being internally dissipated, then P_{dn} is negative and is given by,

$$P_{dn} = -Re\{I_{dn} V_{dn}^{*}\}/2, \; n = \text{output harmonic} \tag{3.33}$$

Therefore, replacing the harmonic voltage $V_{dn} = -I_{dn}Z_L(n\omega)$, for $n > 1$, and assuming the load impedance is conjugate symmetric, power can be represented by current and load impedance instead of current and voltage.

$$P_{\mathrm{DISS}} = I_{D0} V_{D0} - \sum_{n=0}^{n=\infty} |I_{dn}|^2 Re\{Z_L(n\omega)\}/2 \tag{3.34}$$

This expression states that the total power dissipated in the device comprises a dc power minus a sum of ac power delivered to the load

$$P_{\mathrm{DISS}} = P_{DC} - \sum_{n=0}^{n=\infty} P_{out} \tag{3.35}$$

where,

$P_{DC} = I_{D0}\,V_{D0}$, dc power

$P_{\text{out}\ n=no} = |I_{dn}|^2 Re\{Z_L(n\omega)\}/2$, power delivered at the n_oth harmonic

$P_{\text{out}\ n\neq no} = 0$, no power is delivered at all other harmonics

3.4 Fundamental Frequency Power in Amplifiers

The maximum output power is obtained when the magnitude of current and voltage are maximized and their phase relation is equal to 180 degrees. That condition is met at low frequencies where there are no parasitics and the load impedance is purely resistive.

$$P_{\text{outn}} = |I_{dn}|\,|V_{dn}|/2| \tag{3.36}$$

$$\angle I_{dn} - \angle V_{dn} = 180°,\ n > 1$$

The maximum drain voltage swing is a physical constraint of the device and is limited on the low side by the turn on voltage, V_k, and on the high side by the breakdown voltage, BV_{DS}. In a real device, the model in Figure 3.1 has to include the built-in voltage of the diodes, so that the device can deliver a drain current greater than I_{DSS}. The maximum current swing is then determined by I_F, which corresponds to the onset of gate conduction.

3.4.1 Class A

In class A power amplifiers, the device bias is set at half the maximum drain voltage and half the maximum current, so that the maximum power is determined on the I–V plane represented in Figure 3.11, by a load line crossing the points (V_k, I_F) and $(2V_{DD} - V_k, 0)$. The optimum drain bias and the maximum RF peak drain voltage is given by

$$V_{d1} = (V_{DD} - V_k) \tag{3.37}$$

The optimum load for a class A power amplifier is the one that maximizes signal swing within the linear regions represented in Figure 3.11. Therefore the optimum load is given by the ratio of the peak voltage to the peak current,

$$R_{\text{Lopt}} = 2(V_{DD} - V_k)/I_F \tag{3.38}$$

If $R_L < R_{\text{Lopt}}$, the current swing is maintained but the voltage swing is reduced. This load introduces clipping on the current waveform by the gate-

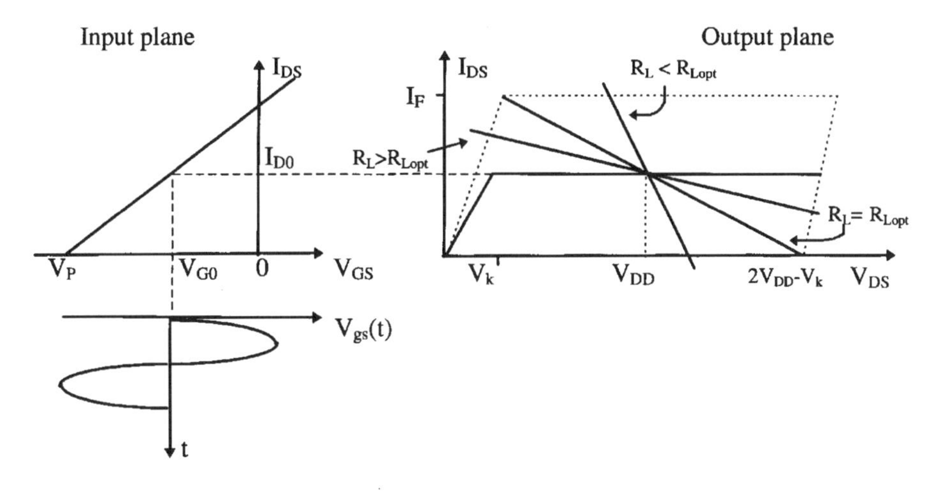

Figure 3.11 Load line for a class A power amplifier.

to-source diode conduction with consequent power loss. The reverse happens if $R_L > R_{\text{Lopt}}$, where the signal is driven into the resistive region and also at the onset of avalanche voltage, BV_{GS}, with similar results. In both cases there is a dc current flowing through either the gate-source or gate-drain diodes. The effect of load-on-drain current and voltage is illustrated in Figure 3.12.

The linear average power delivered by the device is obtained from the product of the rms voltage and current at the fundamental frequency, as given by

$$P_{\text{outA}} = \frac{V_{d1}}{\sqrt{2}} \frac{I_{d1}}{\sqrt{2}} \tag{3.39}$$

The maximum output power is reached for a drain voltage equal to $V_{d1} = (V_{DD} - V_k)$ and a drain current equal to $I_{d1} = I_F/2$. Applying the

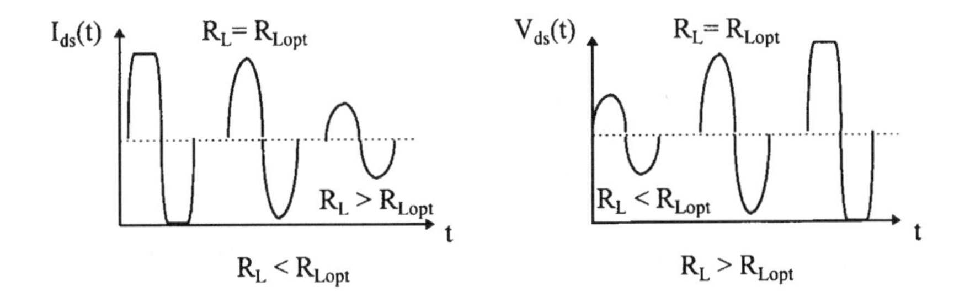

Figure 3.12 Effect of load-on-drain current and voltage waveforms.

parameters of the reference device, $V_{DD} = 4$ V, $V_k \approx 1.0$ V, and $I_F = 90$ mA, one obtains a low frequency $P_{\text{outA}} = 67.5$ mW or 18.3 dBm. The dissipated dc power in the device is given by the difference between dc power and RF power, which is given by (3.40) for the maximum output power condition. In class A, the maximum RF power and the dissipated power in the device are the same.

$$P_{\text{DISS}} = P_{DC} - P_{\text{outA}} = I_F(V_{DD} - V_k)/2 - I_F(V_{DD} - V_k)/4 \quad (3.40)$$
$$= I_F(V_{DD} - V_k)/4$$

The drain efficiency is defined by the ratio P_{outA}/P_{DC}, which in this case is equal to

$$\eta_A = \frac{I_{d1} V_{d1}}{I_F(V_{DD} - V_k)} \quad (3.41)$$

The equation shows that under small-signal operation, V_{d1} is small resulting in low efficiency. Maximum efficiency is obtained when the signal voltage and current are equal to the maximum drain voltage and drain current, and is equal to $\eta_A = 50\%$. If the device is biased such that maximum voltage swing reaches breakdown, then the maximum voltage $(2V_{DD} - V_k)$ can be replaced by $(BV_{DS} + V_P - V_k)$. In such a case, the maximum linear power is defined by

$$P_{\text{outA}} = \frac{(BV_{DS} + V_P - V_k)}{2\sqrt{2}} \frac{I_F}{2\sqrt{2}} = \frac{(BV_{DS} + V_P - V_k)I_F}{8} \quad (3.42)$$

Therefore, to extract maximum power from a given device, one has to drive it at its physical limits in current and voltage. The class A power amplifier just discussed is considered a resistive load line, that is, a load that is resistive at the desired fundamental frequency and harmonics. The tuned power, on the other hand, considers a resistive load only at the output harmonic and a short-circuit at all other harmonics. Thus, a class A power amplifier operation, either at a tuned load or a resistive load, provides the same power since there is a low degree of harmonics generated by the device.

$Z_L(n\omega) = R_{\text{Lopt}}$, for $n \geq 1$ — Resistive load

$Z_L(n\omega) = R_{\text{Lopt}}$, at the output harmonic — Tuned load

$= 0$, for all other harmonics

3.4.2 Class B

In class B, the I–V plot is different and depends on type of load, if resistive or tuned. The resistive load does not present interest in communication circuits and is not considered in this context. A typical high-frequency class B power amplifier circuit with a tuned load is represented in Figure 3.3. The resonant circuit is tuned at the fundamental frequency and all harmonics are shorted. The current and voltage waveforms are indicated in Figure 3.13, where a load line is different from class A operation. Instead of "load line," the world signal path [2] or trajectory [7] in the output plane seems more adequate and is used in this book. The figure also depicts a few instants of time during one cycle, to illustrate the operation of a class B amplifier. At $t = t_1$, the gate voltage is crossing V_P in the direction of maximum gate voltage, 0.7 V. At the same time, the drain current goes from nearly zero at the quiescent point to the maximum value, I_F, at $t = t_2$. After reaching the peak value, the gate voltage returns to the bias point at $t = t_3$, and the drain current and voltage returns to the bias point at the same time. The gate voltage crosses the bias point in direction of maximum negative voltage at $t = t_4$. During the interval $t_3 - t_5$, the drain current is zero, but the drain voltage continues and is sinusoidal due to the filtering action of the tank circuit.

The dc voltage and fundamental frequency voltage component are given by (3.43) and (3.44), which are helpful for calculating the RF power and dc power in the device.

$$V_{D0} = V_{DD} \tag{3.43}$$

$$V_{d1} = (V_{DD} - V_k) \tag{3.44}$$

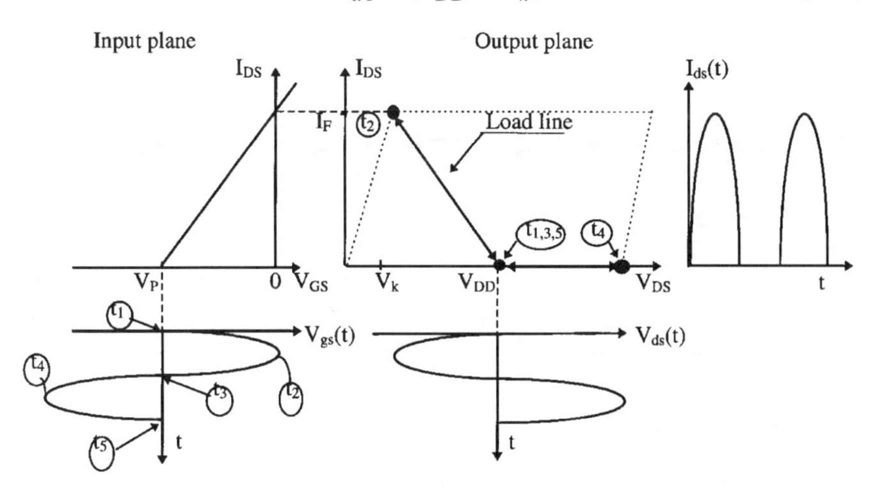

Figure 3.13 Waveforms and load line for a tuned class B power amplifier.

Based on these results, one concludes that the maximum fundamental frequency power in class B is equal to the maximum power in class A. However, the power dissipated by the device is much less, which can be observed comparing (3.46) and (3.47).

$$P_{\text{outA}} = P_{\text{outB}} = (V_{DD} - V_k) I_F / 4 \tag{3.45}$$

$$\text{Class A: } P_{\text{DISS}} = \frac{I_F(V_{DD} - V_k)}{2} - \frac{I_F(V_{DD} - V_k)}{4} = \frac{I_F(V_{DD} - V_k)}{4} \tag{3.46}$$

$$\text{Class B: } P_{\text{DISS}} = \frac{I_F(V_{DD} - V_k)}{\pi} - \frac{I_F(V_{DD} - V_k)}{4} = \frac{I_F(0.273 V_{DD} - V_k)}{4} \tag{3.47}$$

Therefore, class B efficiency, η_B, is higher than class A. For maximum output power ($I_{d1} \approx I_F$) and a low value of saturation voltage ($V_k \approx 0$), $\eta_B = 78.5\%$ which represents a 28.5% increase in efficiency over class A operation.

$$\eta_B = \frac{\pi}{4} \frac{I_{d1}}{I_F} \frac{(V_{DD} - V_k)}{V_{DD}} \tag{3.48}$$

The fundamental frequency impedance is obtained by dividing the voltage component by the current component at the same frequency. The result is equal optimum load impedances for both classes of A and B operation.

3.5 Harmonic Power and Bias in Frequency Multipliers

In frequency multipliers, the objective is to generate a distorted drain current or voltage waveform by means of proper bias and drain termination. Inspection of the I–V plots for the FET represented in Appendix A shows three different biasing regions adequate for generating harmonics, namely pinch-off bias, I_{DSS} or zero gate bias and between both conditions.

3.5.1 Region I: Pinch-off Bias/Class B Multiplier

The proper bias point for a class B multiplier is in the vicinity of pinch-off which generates a half-wave sinusoidal current rich in even harmonics. The circuit schematic of Figure 3.3 represents this condition of operation by making the resonant circuit tuned at the output harmonic. The proper drain termination is the one that induces the highest peak current. That is obtained by a short-circuit at the fundamental frequency indicated in Figure 3.14 by the vertical line crossing the bias point V_{DD}. Ideally, the load should be a short circuit at all harmonic frequencies; however, the presence of a load at a specific harmonic deviates the signal trajectory in the output plane and the load line becomes a function of frequency.

The plot of Figure 3.14 illustrates the case of a frequency doubler, where a few instants of time are selected to explain the multiplying effect. At $t = t_1$, the gate voltage is crossing V_P while the drain voltage is crossing the point $[(2V_{DD} - V_k), 0]$. The gate voltage proceeds from t_1 to t_2, while the drain crosses the entire active area up to the maximum current point, (V_k, I_F) at $t = t_2$. From t_2 to t_3, the gate and drain voltage return to the previous point $[(2V_{DD} - V_k), 0]$. The gate voltage proceeds deep into pinch-off from t_3 to t_4 and back to quiescent point at $t = t_5$. During this time, the drain current is zero, but the drain voltage is not due to the filter action. The device delivers an output voltage at a frequency double than the one applied to the gate.

In this condition of operation, the nonlinear transconductance is the most important harmonic generator. The dc and second harmonic component

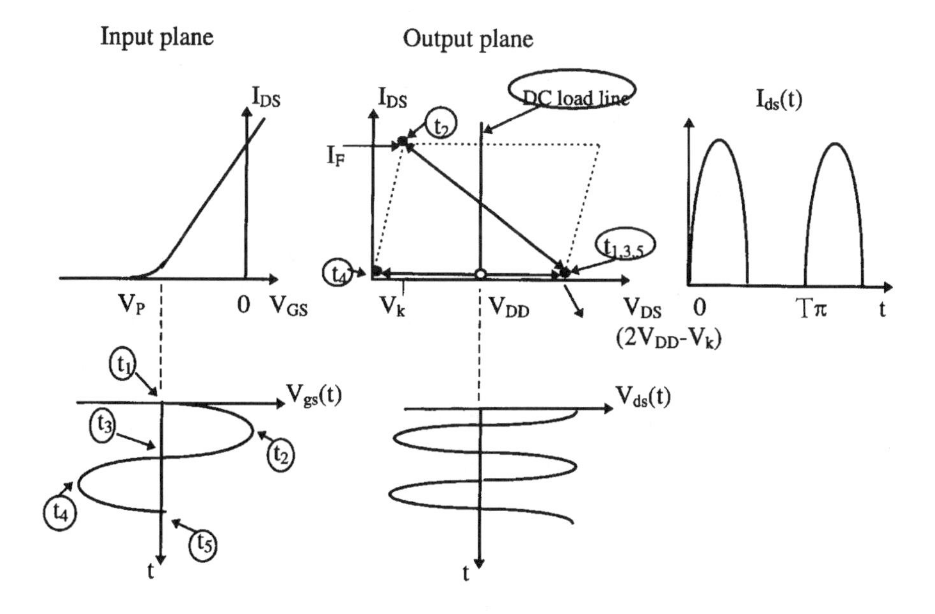

Figure 3.14 Waveforms and signal trajectory for class B multiplier.

of drain-voltage waveform are similar to the dc and fundamental frequency component of the class B amplifier and are given by (3.49) and (3.50). The optimum impedance at the output harmonic is given by the relation between the drain voltage component, and the output harmonic current component.

$$R_{\text{Lopt}}(n\omega) = (V_{DD} - V_k)/I_{dn} \tag{3.49}$$

The harmonic output power is given by the product of the harmonic current component and the voltage component, both taken as rms values for average power calculation.

$$P_{\text{outn}} = \frac{(V_{DD} - V_k)}{\sqrt{2}} \frac{I_{dn}}{\sqrt{2}} = \frac{(V_{DD} - V_k) I_{dn}}{2} \tag{3.50}$$

Comparing the maximum power obtained at the second harmonic with the maximum power obtained at the fundamental frequency by a class B amplifier, the frequency doubler power, P_{outM}, provides 2.3 times less power, or nearly 4 dB less power.

$$P_{\text{outM}} = \frac{(V_{DD} - V_k) I_F}{3\pi} = \frac{4P_{\text{outB}}}{3\pi} \tag{3.51}$$

As an example of the application of this equation, the maximum second harmonic power that can be provided by the reference device was calculated. The output power is equal to $P_{\text{out}} = +14.6$ dBm.

3.5.2 Region II: I_{DSS} Bias

The second important bias point is with a zero gate volt, which in some publications [5] is also referred to as a class A multiplier. This bias point should give the same performance as the pinch-off, if the gate voltage swings from zero to pinch-off volts and a low impedance is connected to the drain. However, a high gate voltage turns the gate-source diode on during part of the positive signal cycle, introducing unacceptable RF losses and a high dc power dissipation. But the device shows a high gain (high transconductance) at this bias point which can be explored by

- Limiting the positive gate-voltage conduction;
- Applying an open circuit to the drain, so that a low-drain current can develop a high-output voltage on the internal impedance.

One way to minimize the gate conduction effect is to apply the dc current, rectified by the gate-source diode, to a large resistor connected from gate-to-ground. Such a current will develop a negative voltage on the resistor terminals which will self-bias the FET proportionally to the signal amplitude. With limited gate conduction, RF losses will be minimal and the operation becomes reliable. The self-bias circuit is represented in Figure 3.15, by the resistor R_B.

The generator is matched by the shunt 50 ohm resistor, and the capacitors C_{bl} and C_b are short circuits at the operating frequencies. The resonant circuit is tuned at the fundamental frequency, thus applying a high impedance to the drain. The inductor L_b is of high value so that it is an open circuit for all harmonics.

The simplified models introduced in Sections 3.1 and 3.2 assumed an infinite output resistance, or zero output conductance. That condition is approximately valid for power amplifiers and class B multipliers, due to the low impedance applied to the drain at the fundamental frequency which shunts the output conductance. However, real models have a finite nonlinear resistance that can be used to generate harmonics, and require an open circuit for its exploitation. Such a termination at the fundamental frequency generates a high-voltage waveform which will be distorted by the output conductance of the device at low-drain voltage. At the other extreme, the drain voltage is high and is determined by the large-signal voltage gain of the device,

$$V_d = -\frac{G_M V_g}{G_{DS} + 1/R_L} \tag{3.52}$$

In this mode of operation, the nonlinear output conductance is the most important harmonic generator, in spite of a significant modulation of transconductance. The drain current and voltage trajectory in the output plane are similar to the one depicted in Figure 3.16 for the case of a frequency doubler. In this plot, consider the five instants of time depicted on the

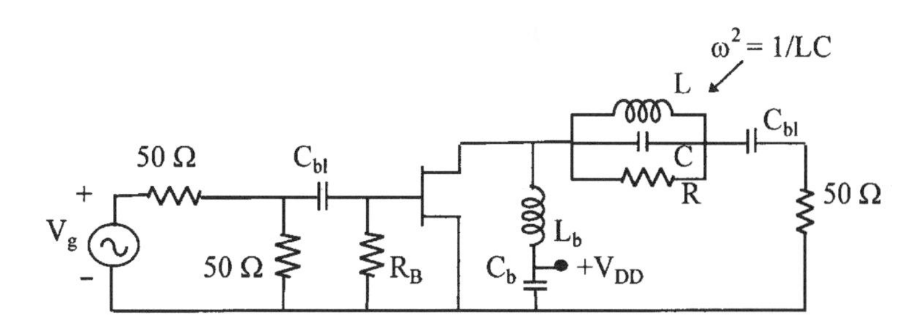

Figure 3.15 I_{DSS} frequency multiplier.

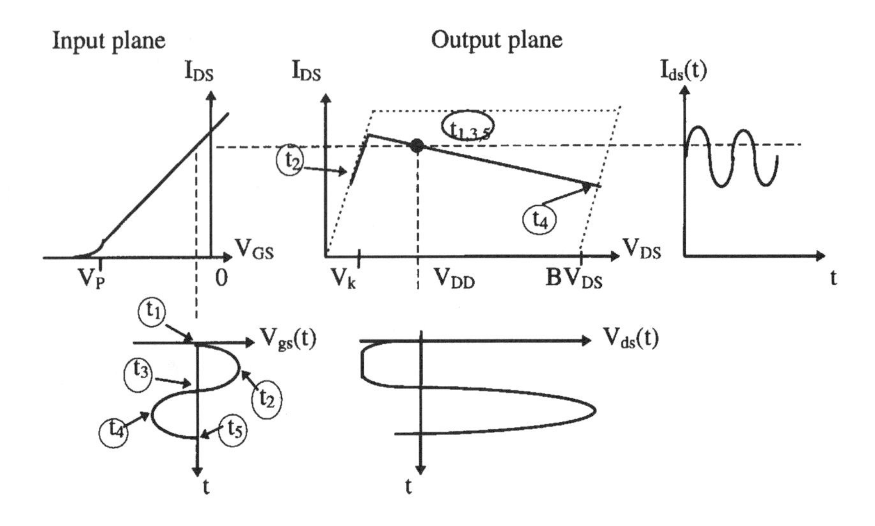

Figure 3.16 Waveforms and signal trajectory for I_{DSS} multiplier.

fundamental frequency gate voltage. At $t = t_1$, the trajectory of current and voltage on the output plane is crossing the bias point in the direction of maximum gate voltage at $t = t_2$. At this instant, the drain voltage enters the resistive region and distorts the drain voltage.

When the gate voltage crosses t_3, the drain voltage crosses the bias point and proceeds to t_4 where the drain voltage amplitude is high and is limited by the output conductance. The resulting drain voltage is distorted and the drain current is sinusoidal due to filter action.

The drain voltage can be approximately described by a train of sine-wave tips whose Fourier coefficients are given by (3.14) to (3.16) with I_p replaced by V_p. The optimum drain voltage is a trade-off between waveform distortion, maximum dynamic drain voltage, and power dissipation. Maximum voltage swing is determined assuming $V_{DD} \approx V_k$, so that the conduction angle becomes close to 180 degrees and the waveform can be approximated by a halfwave cosine. If the gain is high enough, the maximum drain voltage is close to BV_{DS}, and the second harmonic voltage coefficient is given by

$$V_{d2} = 2(BV_{DS} - V_k)/3\pi \tag{3.53}$$

Maximum current swing is obtained when the voltage developed on the self-bias resistor is such that the bias current is $I_F/2$. Thus, the maximum output harmonic power under these ideal conditions is given by

$$P_{out} = \frac{2(BV_{DS} - V_k)}{3\pi\sqrt{2}} \frac{I_F}{2\sqrt{2}} \quad (3.54)$$

Calculating the maximum second harmonic output power for this class of operation, it is found P_{out} = +14.6 dBm, which is the same power provided by the previous bias. However, in practical circuits, these ideal conditions are difficult to meet while, in the pinch-off case, the theoretical conditions are relatively straight forward. One should expect a higher gain and lower power output for this bias condition compared with the pinch-off. The optimum second harmonic load impedance is still given by the relation between harmonic-voltage component and the drain-current component. Assuming full-current swing from zero to I_F, one obtains the result given by

$$R_{Lopt} = \frac{2V_{ds}}{I_F} = \frac{4(BV_{DS} - V_k)}{3\pi I_F} \quad (3.55)$$

3.5.3 Region III: Class A Multiplier

In this region, both the transconductance and output conductance are important nonlinear generators. The device is biased in class A, and the objective is to generate at the output a distorted waveform (current and/or voltage). The signal trajectory will be dependent on the type of load connected to the drain at the fundamental frequency, and on the signal drive. Let us assume a filter is connected to the drain to block the fundamental frequency component, so that the load is reactive. Applying a small amplitude voltage to the gate, in the manner shown in the plot of Figure 3.17, the trajectory on the output plane is an ellipsoid due to the effect of reactive load connected to the drain. Increasing the drive, the extremes of the ellipsoid become distorted, the result of clipping on drain voltage and drain current. This "symmetrical" distortion should approximate a square wave with large-signal drive. Therefore, this bias is adequate to generate odd harmonics at the cost of a high-input power to drive the device into saturation.

Observe that in the figure, drain bias was reduced so that a lower drive power can symmetrically distort the output voltage. The waveforms in the circuit are adequate for odd harmonics generation. The output power in this case is ideally obtained by calculating the third harmonic component of both voltage and drain waveforms, which are symmetrically clipped sinusoid.

$$P_{\text{out}} = C_{\alpha}^{2}[(V_{DD} - V_k)I_P]/2 \quad (3.56)$$

where

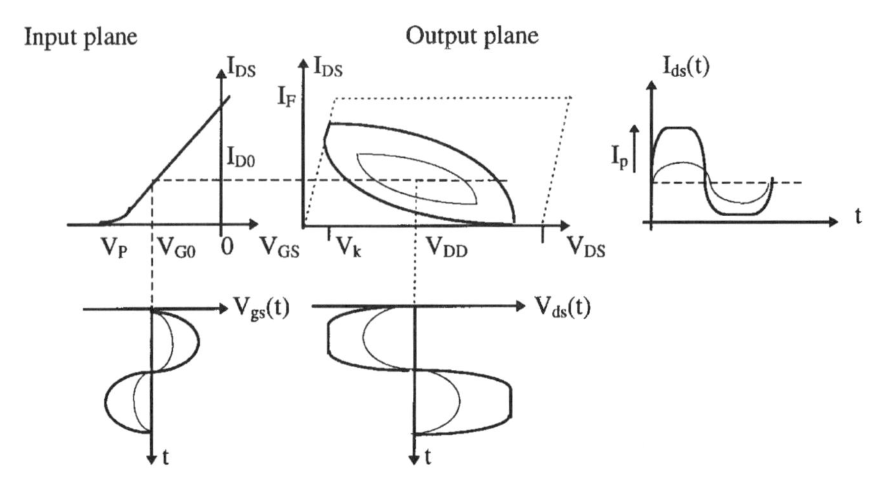

Figure 3.17 Class A multiplier signal trajectory and waveforms.

$$C_\alpha = 2/(3\pi) \qquad \text{for square wave}$$

$$C_\alpha = 2\left[\frac{\tau_0 + \tau_1}{T}\right] \frac{\sin(n\pi\tau_1/T)}{n\pi\tau_1/T} \frac{\sin[n\pi(\tau_0 + \tau_1)/T]}{n\pi(\tau_0 + \tau_1)/T} \qquad \text{for trapezoidal wave}$$

The output power is dependent on output waveform and can range from 4.5 dBm in the case of trapezoidal waveform with 0.5 duty cycle, to 10.8 dBm in the case of a square waveform. The optimum load at the third harmonic will be given by the ratio of voltage and current harmonic components

$$R_{Lopt} = V_{dn}/I_{dn} = (V_{DD} - V_k)/I_p \tag{3.57}$$

References

[1] Fink, G., *Electronics Engineers' Handbook,* New York: McGraw-Hill Book Company, 1975, pp. 2–8.

[2] Clarke, K. K., and D. T. Hess, *Communication Circuits: Analysis and Design,* Malabar, Florida: Krieger Publishing Co, 1994.

[3] Perlow, S. M., "Basic Distortion and Gain Saturation," *Applied Microwave,* May 1989, pp 107–117.

[4] Kushner, L. J., "Output Performance of Idealized Microwave Power Amplifiers." *Microwave Journal,* October 1989, Vol No. 10, pp 103–116.

[5] Carlson, B., *Communication Systems,* New York: McGraw-Hill Book Company, 1968.

[6] Maas, S., *Nonlinear Microwave Circuits,* Norwood, MA: Artech House, 1988.

[7] Rauscher, C., "High Frequency Doubler Operation GaAs Field Effect Transistors," *IEEE Trans. On Microwave Theory and Techniques,* Vol. 31, No. 8, June 1983, pp 462–473.

4

High Frequency Multipliers

The extensive treatment of low frequency multipliers in the previous chapter focused on the direct generation of harmonics from a distorted waveform and is the basis for building frequency multipliers. The analysis made use of a simple model containing the main nonlinear resistive elements within a MESFET. At microwave frequencies, the parasitics associated with the device can no longer be neglected as they modify input/output impedances and introduce unavoidable series and parallel feedback. External feedback can also be used to enhance harmonic generation. Therefore, the generic representation of a frequency multiplier depicted in Figure 4.1 should include not only the input and output networks but the feedback as well.

The generation of even harmonics at low frequencies is obtained by asymmetrically distorting the drain waveform (current or voltage). The odd harmonic generation is obtained by symmetrically distorting the drain wave-

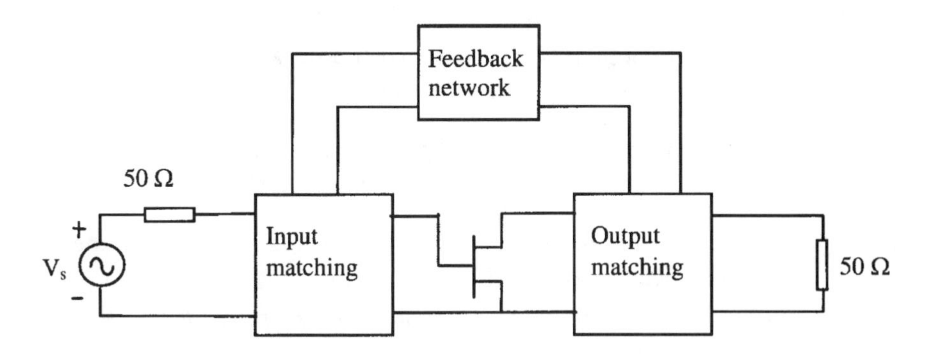

Figure 4.1 Generic high frequency multiplier.

form. It has been shown that the control of such waveforms is carried out by proper bias, drive level, and drain terminations. At high frequencies, the generation of harmonics is provided by the same nonlinear elements, which now are surrounded by parasitic capacitances, inductances, and transmission lines. One can also describe the generation of harmonics as originating not only from current or voltage clipping, but also due to the mixing of fundamental frequency and any one of the generated harmonics. One way to use mixing is to reflect all generated harmonics back to the drain and the other is to feed them back to the gate. Use is made in the first procedure of the output conductance nonlinearity, and in the second procedure, of transconductance nonlinearity [1].

Assuming the harmonics fed back to the input are of a lower level than the fundamental signal power, P_o, the following mechanisms contribute to enhance the output power at the desired harmonic: power amplification of a sample of the output harmonic (n) and frequency conversion by mixing the $(n - 1)$ and/or $(n + 1)$ output harmonic with the fundamental frequency. In an ideal situation where the phase of the feedback signals are such that all power terms add up at the output, then the power at the desired nth harmonic, P_n, can be represented by

$$P_n = (1 - A_n)[P_o MG + P_n A_n G_n + (P_{n-1} A_{n-1} + P_{n+1} A_{n+1}) CG] + \text{other terms} \quad (4.1)$$

where,

P_n = Output power at nth harmonic

P_o = Output power at fundamental frequency, f_o

P_{n-1}, P_{n+1} = Power at the harmonics $(n - 1)$, $(n + 1)$

A_{n-1}, A_{n+1} = Power coupling or reflection coefficient at the harmonics $(n - 1)$, $(n + 1)$

In this equation one can identify the three most important mechanisms contributing to the global multiplication efficiency. They are:

1. Frequency multiplication from the fundamental frequency to the nth harmonic, whose power gain is denoted by MG;
2. Large-signal amplification of part of the nth harmonic fed back to the input with a power gain G_n;
3. Frequency conversion resulting from mixing the fundamental frequency with the feedback harmonics $(n - 1)$ and $(n + 1)$ with a conversion gain denoted by CG.

To take advantage of the output conductance nonlinearity in the mixing process, the drain termination at the fundamental frequency and at the neighboring harmonics $(n - 1)$ and $(n + 1)$ must be such that the power contained at those harmonics is reflected back to the device with an adequate phase to maximize the voltage amplitude at the drain, optimizing the mixing with the fundamental frequency component. These conditions are difficult to implement at high frequencies due to the presence of parasitics that present high reactances at the harmonic frequencies. The application of nonlinear transconductance to enhance harmonics in this process is even more complex since it requires feeding back to the gate the harmonics $(n - 1)$ and $(n + 1)$, with control of loop gain under unity to maintain stability. The objective of this chapter is to identify the high-frequency effects on frequency multipliers, and investigate how each of the described effects can be used to improve a multiplier performance.

4.1 The High Frequency Model

The high-frequency model, represented in Figure 4.2, includes the nonlinear gate-source and gate-drain capacitances, which in this model are approximated by the capacitance of reverse-biased Schottky-barrier diodes. The model includes the parasitic intrinsic capacitances C_{ds}, the bonding pad capacitances C_{gp}, and a parallel feedback capacitance C_{gd2}. The wire bond inductances L_g, L_s, L_d are

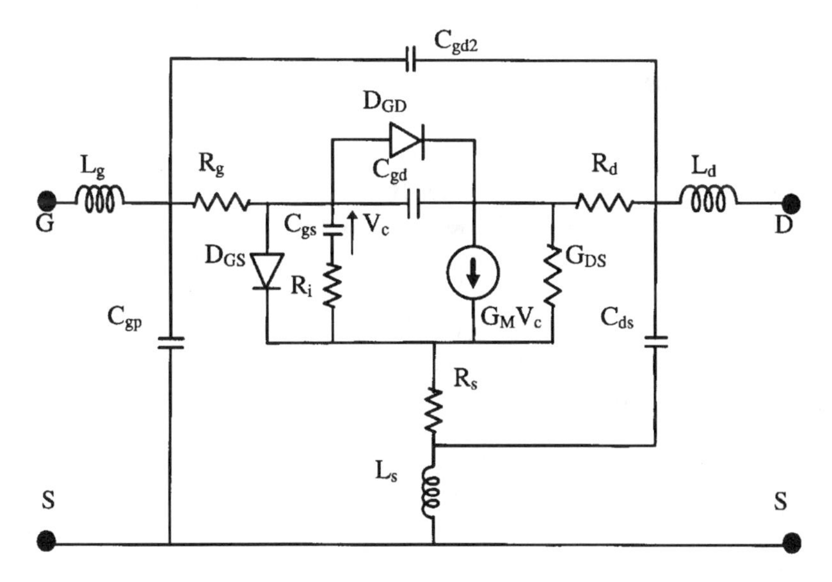

Figure 4.2 High frequency nonlinear model.

not from the device itself but they are unavoidable when the die is connected to external circuitry.

It is obvious from this figure that the number of variables involved in a high-frequency multiplier is so high that some limitations will have to be introduced to obtain conclusions that are readily applicable to practical circuit design. Therefore, let us eliminate the bonding pad and feedback capacitances and use the resulting circuit for a simplified high frequency analysis.

The first high-frequency effect to take into account the design is the bias selection which now has to consider the fact that maximum voltages may be present simultaneously at the input and output of the FET. This condition may happen at high frequencies due to phase shift introduced by the FET and due to the impedance of the terminations connected to the gate and drain. Thus, if the gate is biased in deep class C ($V_{G0} \geq V_P$) with the purpose of decreasing the duty cycle and increasing the desired harmonic amplitude, the reverse voltage on the gate-drain diode may exceed the device avalanche voltage. This extreme condition may happen if the amplitude of the gate-source voltage is equal to ($V_{gmax} - V_{G0}$), and the reverse voltage is equal to $[V_{dmax} - (V_{gmax} - V_{G0})]$. Thus, for practical high-frequency applications, gate bias should be limited to class A, class AB, or class B operation. If class C is required in order to obtain a duty cycle less than 50%, then a careful computer simulation should be done in order to determine signal phase shift and to trade-off bias and signal amplitude for the best harmonic generation.

4.2 Drain Effects

An important high frequency effect is noticed at the drain terminal where the nonlinear elements are no longer directly accessible. Thus, in the case of class B multipliers, a short circuit between drain and source terminals does not reproduce the same low frequency condition close to the internal nonlinear elements. This can be visualized in a more effective way if the circuit of Figure 4.2 is simplified to represent only the drain circuit, shown in Figure 4.3. The nonlinear elements are the current source, $I_{dn} = G_M V_g$, and the output conductance, G_{DS}, which are separated from the output short by ($R_s + R_d$), ($C_{ds} + C_{gd}$), ($L_s + L_d$) elements. As a matter of fact, the drain load at the fundamental frequency has to series resonate the drain-plus-source inductances, to approximately obtain the same low frequency effect.

The low drain impedance at the fundamental frequency is in shunt with the output conductance, minimizing its nonlinear effects on the overall performance, and making the transconductance the most important nonlinear

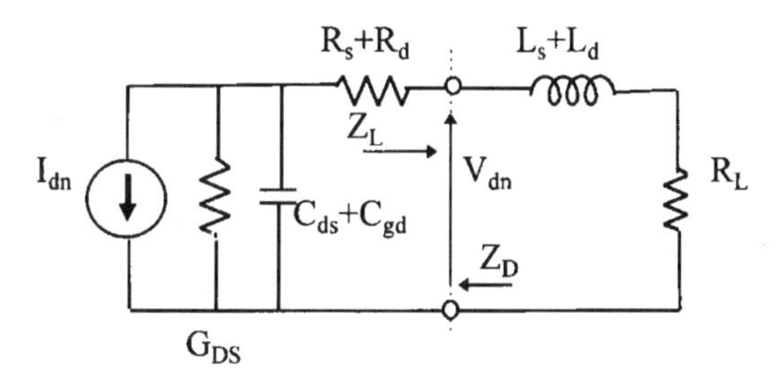

Figure 4.3 Simplified high frequency drain circuit.

element. The output harmonic voltage at the device terminal, V_{dn}, is given by

$$V_{dn} = \frac{I_{dn}}{(G_{DS} + j\omega C)} \frac{(R_L + j\omega L)}{[R_L + j\omega L + R + 1/(G_{DS} + j\omega C)]} \tag{4.2}$$

where,

$$R = R_d + R_s$$
$$C = C_{ds} + C_{gd}$$
$$L = L_s + L_d$$

The previous equation assumes a constant current source at the output harmonic driving a complex load. Resonating the load reactance at the output harmonic, higher voltage is available which translates into higher output power. Power can be increased further if the load resistor is also increased. However, there is a limitation on the magnitude of the harmonic drain voltage, restricted by the device internal impedance. The load impedance is given by $Z_L = R_L + j\omega L$, the device output impedance given by $Z_D = R + 1/(G_{DS} + j\omega C)$, and the output power at harmonic n can be expressed by

$$P_n = \frac{|I_{dn}|^2}{2|G_{DS} + j\omega C|^2} \frac{Re(Z_L)}{|Z_L + Z_D|^2} \tag{4.3}$$

The matched conditions are obtained when $Z_L = Z_D^*$, which in this case is also the condition for maximum power, as expressed by

$$P_n = \frac{|I_{dn}|^2}{8|G_{DS} + j\omega C|^2} \frac{1}{Re(Z_D)} \tag{4.4}$$

Keep in mind that the power conditions are dependent on the fundamental frequency termination which has a direct impact on I_{dn}. Thus, the matching at the output harmonic can be applied if they do not interfere with the fundamental frequency termination. A similar result is obtained for the class A or I_{DSS} multiplier, where a distorted voltage waveform is established at the drain. In that case, the high-load impedance connected to the drain has to parallel resonate the bonding wires plus capacitances to obtain a high voltage swing at the internal drain impedance at the fundamental frequency. Therefore, the current generator will develop a distorted voltage at its internal nonlinear output conductance and both elements will generate harmonics.

The equivalent harmonic voltage generator represented in Figure 4.4 has an associated equivalent internal resistance, $R_{DS} = 1/G_{DS}$, obtained from the average value of the output conductance under large-signal conditions. Therefore, at the output harmonic, the load also has to be able to resonate the bonding wire inductance and the equivalent capacitance to obtain maximum efficiency. The equation for output power at the output harmonic is similar to the one employed in the previous case and is given by

$$P_n = \frac{|V_{dn}|^2}{2|1 + j\omega CR_{DS}|^2} \frac{Re(Z_L)}{|Z_L + Z_D|^2} \tag{4.5}$$

4.3 Gate Effects

The next high-frequency effect is observed at the input, where the voltage from the generator is no longer directly applied to the control capacitor, but to a complex gate impedance. At high frequencies, a high standing wave may be established with very low voltage developed on the control capacitor, C_{gs}, resulting in poor multiplication efficiency. One way to maximize that voltage is to apply the principles of maximum energy transfer from a source to a load. This condition is accomplished in a manner similar to amplifier design, where

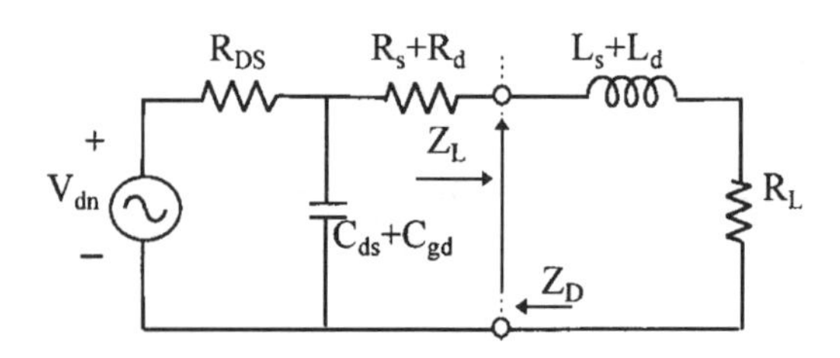

Figure 4.4 Equivalent voltage harmonic generator.

the input impedance is conjugately matched to the generator impedance if the device is stable.

The problem then is to model an equivalent input impedance, dependent on load termination and on signal level, to design the matching network. When deriving the equivalent input impedance, one must also take into account the input capacitance nonlinearity at high frequencies. An accurate equivalent impedance can only be obtained from a full nonlinear simulation. However, approximate values can be determined from the following analysis which assumes the gate-to-source capacitance, C_{gs}, varies nonlinearly with the gate voltage and is independent of drain voltage.

A reasonable approximation to simulate this capacitance is to employ (2.2) and (2.3) to describe its value within the active region and in the pinch-off region as a function of voltage. There are two ways to obtain an equivalent value for the nonlinear capacitance: (1) the small-signal approximation which is applicable to most class A amplifiers; and (2) the large-signal approximation applicable to other nonlinear circuits.

4.3.1 Nonlinear Reactance/Small-signal Approximation

In this condition, the square root capacitance-voltage relation can be approximated by a Taylor series. This approximation is described by (4.6), where the terms above the second were neglected.

$$C(V_{gs}(t)) = C_{GS0}\left[1 + \frac{1}{2}\frac{V_{gs}(t)}{(V\phi - V_{G0})} + \frac{1}{4}\frac{V_{gs}(t)^2}{(V\phi - V_{G0})^2}\right] \tag{4.6}$$

where,

C_{GS0} = capacitance at the bias point V_{G0}

$V_{gs}(t) = V_g \cos\omega t$, AC voltage at the capacitor terminals

This equation shows that at low signal levels the ac terms are very small, and the equivalent capacitance is equal to the capacitance of the bias point. When signal amplitude is increased, the capacitance at the bias point will alternate proportional to the amplitude of applied signal. However, the average or equivalent value is still the same, (i.e., the capacitance at the bias point). Increasing the signal amplitude even more, the square law term becomes important and total capacitance is found by replacing $V_{gs}(t)$ by the cosine term. This description is valid as long as the applied voltages are much lower than the applied bias. Therefore, the total capacitance becomes proportional to the square of signal amplitude.

$$C(V_{gs}(t)) = C_{GS0}[1 + kV_g^2] \tag{4.7}$$

where,

$$k = \frac{1}{8}\frac{1}{(V\phi - V_{G0})^2}$$

4.3.2 Nonlinear Reactance/Large-signal Model

Application of Taylor series is only practical at low levels due to the high number of terms required to represent a nonlinearity under large-signal conditions. In such cases a Fourier analysis is more accurate and is used to determine the equivalent value of nonlinear reactance at the fundamental frequency. The equivalent capacitance is obtained by averaging the charge variation over one cycle of applied voltage. The capacitance is defined as the derivative of charge with respect to voltage

$$C(V) = dQ(V)/dV \tag{4.8}$$

where $Q(V)$ is the charge on the capacitor terminals. The charge is obtained by integrating the capacitance with respect to voltage.

$$Q(V) = \int C(V)\, dV \tag{4.9}$$

The nonlinear capacitance voltage relationship can also be expanded by a Fourier series as

$$C(V) = \sum_{n=0}^{n=\infty} C_n e^{jn\omega t} = C_{GS0}(V_{G0}) + \sum_{n=1}^{n=\infty} C_n e^{jn\omega t} \tag{4.10}$$

The FET input matching circuit is mainly concerned with the fundamental frequency, so that only the average value of the series needs to be considered. Assuming a cosine voltage $V_{gs}(t) = V_{G0} + V_g \cos\omega t$ is applied to the capacitor, after integrating (4.9) the charge is described as

$$Q(V_{gs}(t)) = C_{GS0}(V_{G0})\, V_g \cos\omega t \tag{4.11}$$

The average capacitance can be obtained by multiplying this equation by $\cos\omega t$ and integrating over one cycle of applied signal, and dividing by the total duration 2π. Therefore,

$$\int_0^{2\pi} Q(V_{gs}(t))\cos(\omega t)d(\omega t) = \int_0^{2\pi} C_{GS0}(V_{G0})\, V_g \cos^2(\omega t)d(\omega t) \qquad (4.12)$$

$$C_{GS0}(V_{G0}) = \frac{1}{\pi V_g}\int_0^{2\pi} Q(V_{gs}(t))\cos(\omega t)d(\omega t) \qquad (4.13)$$

Based on this equation, the equivalent capacitance is obtained from the charge as a function of applied voltage, which requires integration of (2.2) and (2.3) within the active and pinch-off regions. The resulting charges, Q_1 and Q_2, are respectively,

$$Q_1(V_{gs}) = -C_{GS0}[V_\phi(V_\phi - V_g)]^{1/2} + q_0 \qquad \text{for } V_P < V_{GS} \qquad (4.14)$$

$$Q_2(V_{gs}) = \frac{V_g C_{GS0}}{\sqrt{1 + V_P/V_\phi}} + q_1 \qquad \text{for } V_P > V_{GS} \qquad (4.15)$$

where, $q_{0,1}$ = integration constant to maintain charge continuity.

A conventional capacitor has no stored charge with zero applied bias. In MESFETs, the zero gate bias does not imply zero bias at the internal capacitors due to the voltage drop at the source resistance and to the built-in voltage of the Schottky-barrier. Therefore, the constants of integration are not zero. The total average charge is given by the average of charges Q_1 and Q_2.

A physical explanation of the dependence of equivalent capacitance as a function of bias voltage and large-signal voltage can be achieved by graphical analysis, employing the $Q(V_{gs})$ plot. Figure 4.5 illustrates the plot of (4.14) and (4.15), which provides a graphical relation between the charge at the gate

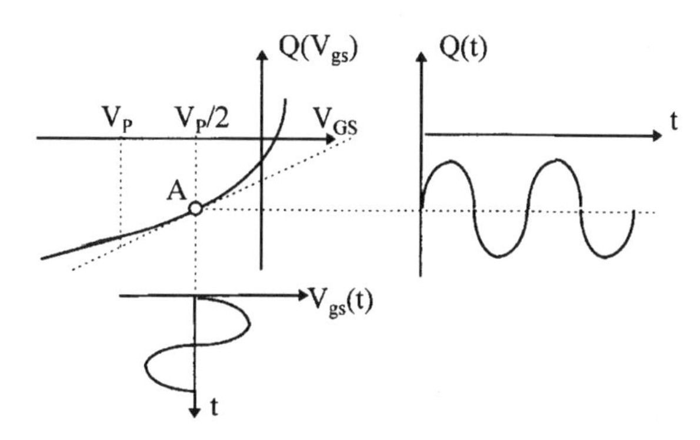

Figure 4.5 Charge modulation for class A bias.

as a function of applied voltage. It also shows the charge modulation due to application of a sinusoidal voltage to the gate of a device biased in class A, $V_{GS} = V_P/2$, indicated by point A on the plot. Observe that the distortion on the charge by the application of a large-signal is low. Thus, the average charge or capacitance is approximately given by the capacitance at the bias point, therefore similar to the small-signal result.

In Figure 4.6, the device is biased in class B, $V_{GS} = V_P$ indicated by point B, and a large-signal voltage is superimposed to the bias. Observe that in the negative side of the applied waveform, $Q(V_{gs})$, is linear and therefore $Q(t)$ is also sinusoidal. On the positive side, $Q(V_{gs})$ is not linear and distorts the sinusoid. From the $Q(t)$ plot, it is obvious that the average value over the signal period is shifted upward to point B′. The increase in capacitance proportional to signal level can still be given by (4.7) with the appropriate constants.

Biasing the device at I_{DSS}, in Figure 4.7, a highly distorted charge is produced. An approximate graphical solution consists of the linearization of the $Q(V_{gs})$ relation in two parts, linear between V_P to zero and between zero to the onset of gate conduction. The average capacitance is then determined from the average of both linear functions. Comparing the two areas of $Q(t)$, (i.e. the one at $V_{GS} > 0$ with the one at $V_{GS} < 0$), in the figure, note that the latter has more area so that the average charge over one period is shifted downward to point C′ lowering the capacitance at the bias point.

An accurate determination of the equivalent capacitance requires integration of (4.13) which is a complex task best left to a computer to do a numerical integration. Due to this difficulty, a simple approximate alternative is to obtain the equivalent capacitance of a MESFET under large-signal conditions at low-frequency, as a function of bias drive voltage. The reason to perform the

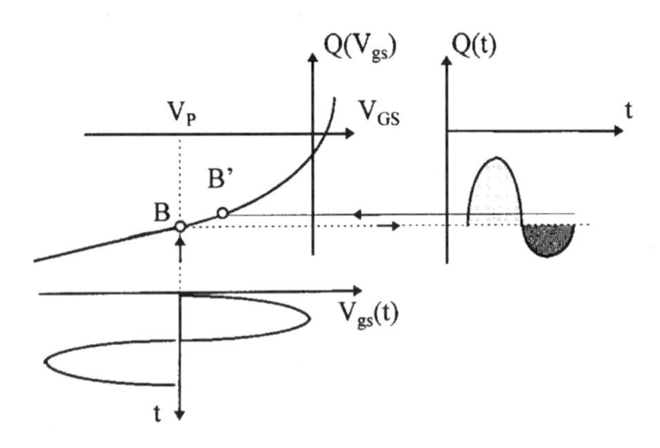

Figure 4.6 Charge modulation for class B bias.

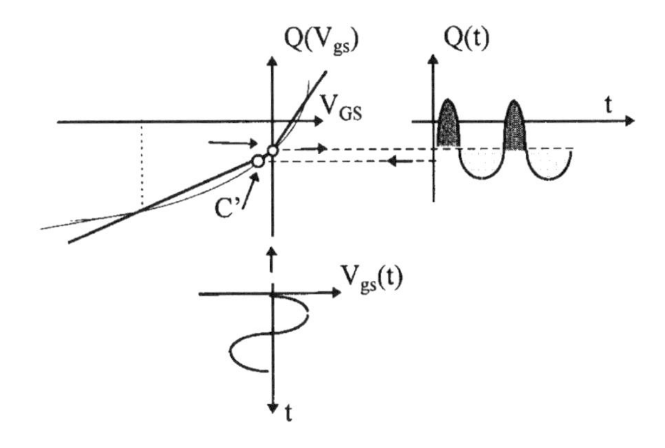

Figure 4.7 Charge modulation at I_{DSS} bias.

simulation at low-frequency is that the generator voltage is directly applied to C_{gs} without the need of a matching circuit. The simple circuit of Figure 4.8 can be simulated by any commercial nonlinear software package to calculate the gate current due to the applied voltage, V_s.

The average gate-source capacitance is obtained from the large-signal input impedance, defined by (4.16). This procedure is valid if caution is taken to maintain at high-frequency the same capacitor voltage used at low frequency.

$$Z_{\text{in}} = V_s / I_g \tag{4.16}$$

After estimating the equivalent value for C_{gs} for a given signal amplitude, it is possible to obtain the high frequency equivalent input impedance from (4.17) and (4.18), for a shorted drain termination. These expressions are valid only if the circuit is active, either class A or class AB bias. Otherwise they will have to be modified. The large-signal transconductance for the linear model of Chapter 3 presents the same small-signal and large-signal transconductance. Therefore, for a class B multiplier, the input impedance is equal to

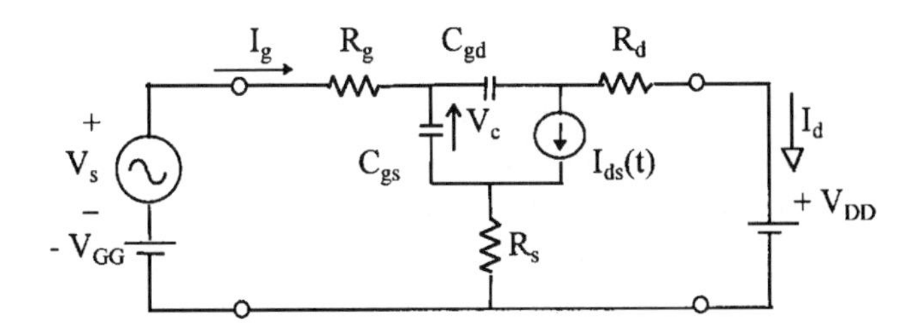

Figure 4.8 Low frequency circuit for determination of input capacitance.

$$Z_{\text{insc}} = R_g + R_s + \frac{G_M L_s}{C_{eq}} + j\left[\omega L_g - \frac{1}{\omega C_{eq}}\right] \tag{4.17}$$

An accurate evaluation of the input impedance can be obtained from S-parameters of the device, as described by

$$S_{11}' = S_{11} + \frac{S_{12} S_{21} \Gamma_L}{1 - S_{22} \Gamma_L} \tag{4.18}$$

where,

$$\Gamma_L = \frac{Z_L - Z_0}{Z_L + Z_0} = \text{Load reflection coefficient} \tag{4.18a}$$

$$S_{11}' = \frac{Z_{\text{in}} - Z_0}{Z_{\text{in}} + Z_0} = \text{Input reflection coefficient when the output is terminated in } \Gamma_L. \tag{4.18b}$$

Use of (4.18) requires consideration of the signal level employed during measurements or simulations, due to its validity only under small and mild signals. However, the reflection coefficient Γ_L and S_{11}' described by (4.18a) and (4.18b) are always valid and can be employed at large signals. If the drain is open-circuited, case of class A or I_{DSS} multiplier, then (4.17) has to be modified to (4.19). The problem in this case is that the output conductance can differ significantly from the small-signal condition and the equation in this case is not of general use. It can however, be used to qualify the dependence of the elements on input impedance. The average values of the nonlinear elements can only be determined through numerical iterative procedures, where an initial value is modified at each simulation until an acceptable convergence is obtained.

$$Z_{\text{inoc}} = R_g + R_s - \frac{G_{DS0}^2}{\omega^2 C_{eq}^2 (G_{M0} + G_{DS0})} + j\left[\omega(L_s + L_g) - \frac{1}{\omega C_{eq}}\right] \tag{4.19}$$

Observe that the open-circuited condition shows the possibility of negative resistance at low frequencies, due to internal positive feedback in the device. Therefore, the drain reactive termination must be selected such that the input impedance is positive for stable operation. Another important high frequency effect is the dissipation of RF power at the gate, which is no longer negligible and can be estimated by

$$P_{\text{in}} = 0.5 Re(Z_{\text{in}})[\omega C_{eq} V_g]^2 \tag{4.20}$$

4.4 Feedback Effects and Stability

By their very nature, microwave transistors are active devices with intrinsic feedback, which may become positive when associated with an external impedance, greatly affecting the circuit stability and the terminal impedances. The stability theory [2] developed for amplifiers is still applicable to analyze this type of topology if the circuit is active, and can be calculated from S-parameters. It is usually divided into two categories:

Unconditionally stable
If $K > 1$ and $B_1 > 0$ and $B_2 > 0$, the active device is considered unconditionally stable for any input and output terminations.
Conditionally stable
If $K < 1$, or $B_1 < 0$ or $B_2 < 0$, the active device is considered conditionally stable. Therefore, stability is conditioned to certain input and output terminations. The more K is less than 1, the more unstable the circuit will be.

The above parameters are defined by the following set of equations

$$K = \frac{1 + |\Delta|^2 - |S_{11}|^2 - |S_{22}|^2}{2|S_{12}S_{21}|} \tag{4.21}$$

$$B_1 = 1 + |S_{11}|^2 - |\Delta|^2 - |S_{22}|^2 \tag{4.22a}$$

$$B_2 = 1 + |S_{22}|^2 - |\Delta|^2 - |S_{11}|^2 \tag{4.22b}$$

where, $\Delta = S_{11}S_{22} - S_{12}S_{21}$.

Microwave devices are unconditionally stable only within certain frequency ranges, usually above a certain minimum frequency. However, frequency multipliers biased class A or AB may operate at a rather low input frequency, below such minimum frequency, so that circuit stability has to be carefully examined. The classical topology for active frequency multipliers is to load the drain with a reactive termination and match the gate at the fundamental frequency. This circuit condition may induce a negative resistance looking at the gate if the device is biased at a specified bias current. The set of drain impedances, expressed by Γ_L, that will induce negative resistance can be determined by making $|S_{11}|' > 1$ in (4.18), and solving for Γ_L. The resulting drain impedances are represented by a circle on the Smith chart defined by

$$\text{center of circle} \quad C_o = C_2^*/(|S_{22}|^2 - |\Delta|^2) \tag{4.23}$$

$$\text{Radius of circle} \quad R_o = |S_{21}S_{12}|/|(|S_{22}|^2 - |\Delta|^2)| \tag{4.24}$$

where, $C_2 = S_{22} - \Delta S_{11}^*$ and the asterisk represents complex conjugate.

In order to find out which impedances causes the input to present negative resistance, one has to terminate the output so that $\Gamma_L = 0$ and check the input reflection coefficient:

If the input impedance is positive and if the circle includes the origin, then the inside of the circle represents the stable region.
If the input impedance is positive and if the circle excludes the origin, then the inside of the circle represents the unstable region.

If the gate is biased in class B or C, then the circuit is not active without RF drive and the circuit is stable. On the other hand, looking into the drain at the fundamental frequency, it is less probable to find a negative resistance if the gate is terminated in a low reactance. However, to guarantee the circuit stability, both the input and output impedances have to be checked for stability. To check the stability on the output plane, it suffices to interchange subscripts 1 by 2, in (4.23) and (4.24).

Therefore, to check the effect of reactive drain termination on the input impedance and the circuit stability at the fundamental frequency, the drain impedance is made equal to $\Gamma_L = 1\angle\theta$, with $0 < \theta < 2\pi$. A numerical example is useful to clarify these statements: The input impedance in terms of S_{11}' and the stability circle for the output plane were calculated from the S-parameters generated for the reference device at 5 GHz biased at a point corresponding to 10% I_{DSS}.

The resulting stability analysis is depicted in the Smith chart plot of Figure 4.9, where circle A, represents the stability circle in the output plane

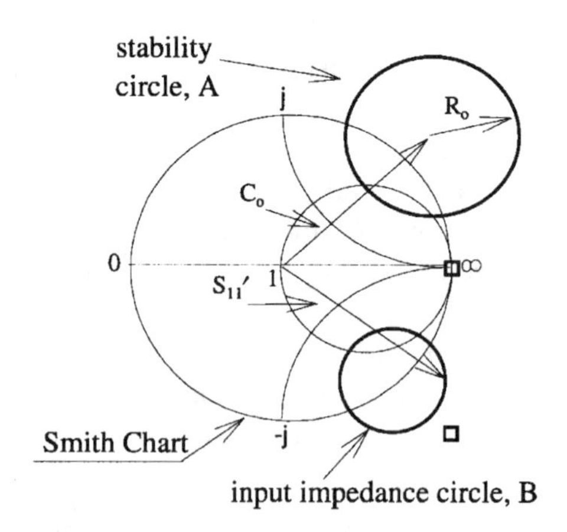

Figure 4.9 Effect of load on stability and input impedance.

crossing regions of negative impedance. The interior of the circle represents the drain impedances that will induce negative resistance at the input. In the same plot there is another circle, B, representing the input impedance for the same range of drain load phase variation. Observe that part of the circle is on the negative resistance area.

Therefore, the plot shows which are the reactances that should be avoided in order to guarantee stability. The represented input impedance still has to be corrected for large-signal operation, a procedure that will be discussed in the next chapter. The stability test has to be repeated over the entire operating frequency band of the circuit.

4.5 Frequency Doubler Case Study

It has been demonstrated through linear analysis that the load termination at the fundamental frequency has a major effect on circuit stability. The same load termination controls the series and parallel resonance of the device parasitics, controlling the peak of rectified current or the distorted drain voltage. Thus, it also has an important effect on large-signal effects such as multiplier gain, input impedance, and bandwidth.

An interesting demonstration [3, 4] of this dependence is described in the literature for the case of a frequency doubler using a linear model for the device. In the circuit of Figure 4.10, the drain circuit has been split into two, one for the fundamental frequency and the other for the second harmonic. The drain termination at the fundamental frequency is a shorted stub with a variable electrical length ϕ, and at the second harmonic, it is assumed to be matched.

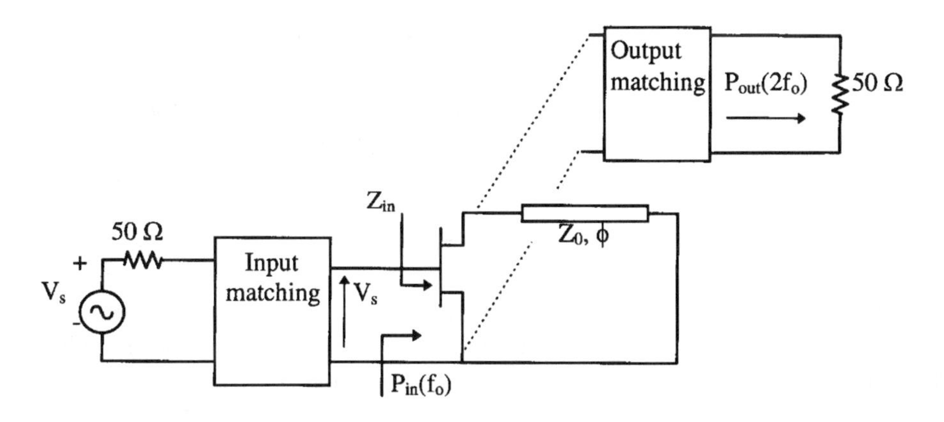

Figure 4.10 Simplified frequency doubler circuit.

According to (3.16), if the device is biased in class B, the second harmonic component is given by $(2I_{\text{peak}})/(3\pi)$, and the peak current is given by (4.24). Taking this relation and the input power defined by relation (4.25) to the multiplication gain equation, defined by (4.26), one obtains:

$$I_{\text{peak}} = I_{DSS}(V_P - V_{G0} + V_g)/V_P \tag{4.25}$$

$$P_{\text{in}}(f_o) = R_{in}|I_g|^2 = \frac{R_{\text{in}}|V_g|^2}{|Z_{\text{in}}|^2} \tag{4.26}$$

$$MG = \frac{P_{d2}(2^{\text{nd}}\text{ harmonic})}{P_{\text{in}}\text{ (fundamental frequency)}} = \frac{R_L|I_{d2}|^2}{P_{\text{in}}} \tag{4.27}$$

$$MG \propto \frac{R_L G_{M0}^2 |Z_{\text{in}}|^2}{R_{\text{in}}} \tag{4.28}$$

This equation shows that the multiplication gain can be roughly determined by calculating the equivalent input impedance of the circuit as a function of the angle of a short-circuited transmission-line, also denominated short-circuited stub.

There are mainly two parameters of a transmission-line, namely, its characteristic impedance, Z_0, and its electrical angle, $\phi°$. Their definition can be found in any text book on transmission-line theory [5].

The characteristic impedance of a lossless transmission line is defined by (4.28) and is obtained from the ratio of the series inductance and parallel capacitance per unit length of a transmission-line structure.

$$Z_0 = \sqrt{L/C}, \text{ in unit of ohms} \tag{4.29}$$

The electrical angle, defined by (4.29), is a function of the transmission-line length, ℓ, and the frequency of operation, expressed in terms of wavelength, λ_e.

$$\phi = (2\pi l)/\lambda_e \tag{4.30}$$

Normalizing the gain to the gain given by a shorted-stub measuring $\phi = 0°$ length, one obtains the multiplication gain for this type of multiplier. The results in Figure 4.11 were obtained for a MESFET cell with $0.5 \times 100\ \mu$m gate area, and calculated for three relevant microwave fundamental frequencies.

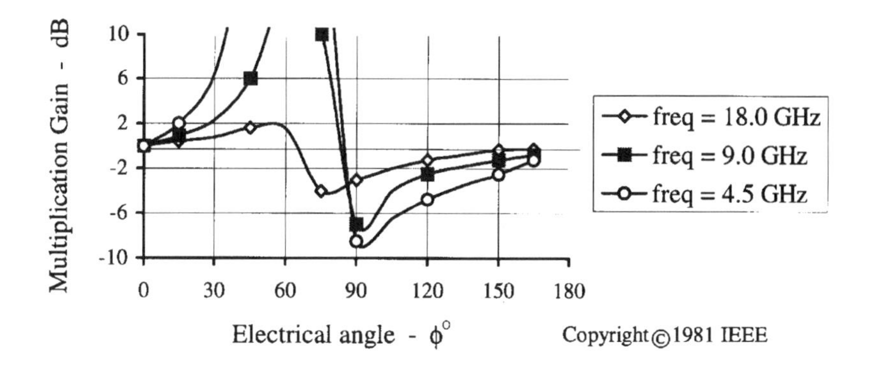

Figure 4.11 Multiplication gain as a function of angle and frequency [4]. (*Copyright 1981 by IEEE.*)

Observe from this plot that at low frequencies there is a set of angles that results in infinite gain, representing a negative resistance effect on input impedance. As frequency increases, the range of angles where the gain is too high narrows and above a certain frequency, the multiplication gain is defined for all angles. A similar result is obtained for the input impedance of a frequency doubler as a function of the phase angle for two frequencies and gate widths dimensions, which are shown in Figure 4.12. The higher resistance and lower reactance curves are for a 100 μm gate-length FET at 12.5 GHz and the other set is for a 200 μm gate-length at 6.5 GHz.

From this plot one can obtain the circuit Q, defined by (4.30) and plotted in Figure 4.13. The Q-factor is undefined in the region of negative resistance.

$$Q = Im[Z_{\text{in}}]/Re[Z_{\text{in}}] \tag{4.31}$$

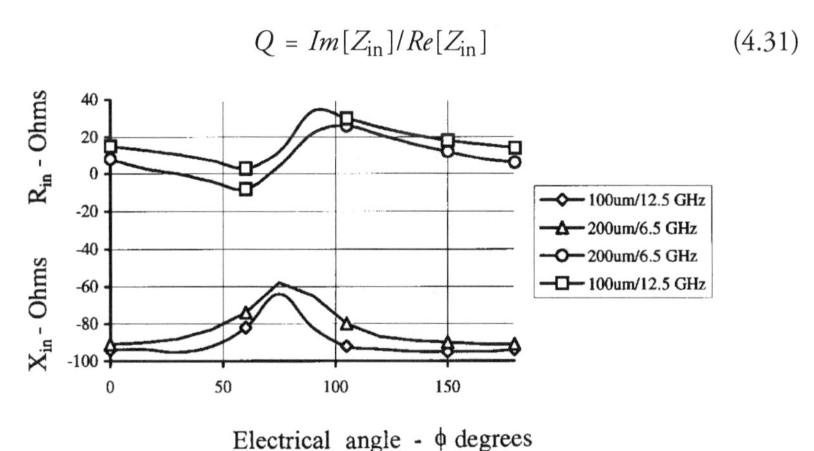

Figure 4.12 Gate impedance as a function of drain reactance phase [4]. (*Copyright 1981 by IEEE.*)

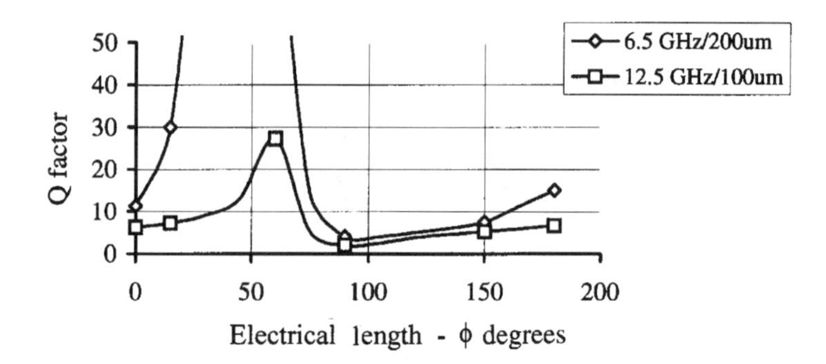

Figure 4.13 Quality factor for the input impedance of reference device [4]. (*Copyright 1981 by IEEE.*)

Observe that the high frequency device, having a shorter gate width, presents a lower Q, enabling large bandwidths to be obtained. The lower frequency presents a higher Q which is minimum either close to the drain or after inserting a certain length of line between the device and the short-circuit position. Therefore, high multiplication gain can be obtained with a higher circuit Q and that implies a narrower band of operation. Note that bandwidth will have to be traded for multiplication gain.

A better insight into the bandwidth as a function of the drain-shorted stub was demonstrated [4] for a L-C input matching circuit for the range $\varphi = 0$ to 50 degrees. The circuit in Figure 4.14 gives a perfect match at the center frequency of operation, with L and C given by (4.30) and (4.31), where all elements are normalized to Z_0.

$$L = \{-X_{in} + [R_{in}(1 - R_{in})]^{1/2}\}/\omega \tag{4.32}$$

$$C = [(1 - R_{in})/R_{in}]^{1/2}/\omega \tag{4.33}$$

The "bandwidth" was considered to be an input impedance providing better than 10 dB return loss as a function of electrical length. The results at

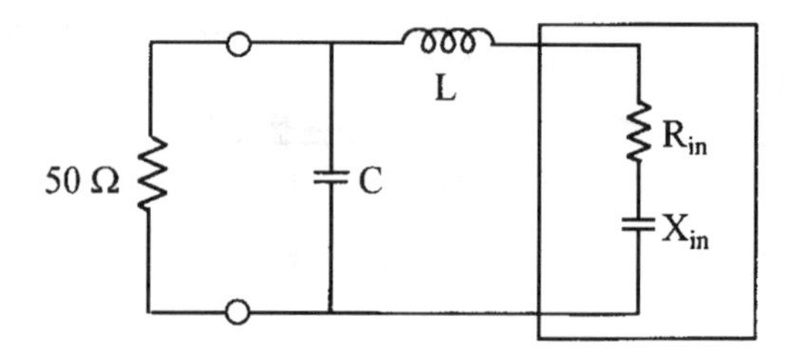

Figure 4.14 Matching circuit for the gate impedance.

12.5 GHz for 100 μm and 200 μm gate-length devices are shown in Figure 4.15 by the top curves. The bottom curves are for the same dimensions but calculated at 6.5 GHz.

The simulation shows that a good trade-off for gain and bandwidth is a very short electrical length. Narrow-gate width gives a better trade-off at the expense of output power. The introduction of external feedback to enhance the multiplier efficiency is an excellent alternative if the device possesses reasonable power gain at the feedback frequency. Such a process is adequate if the order of multiplication ratio is high, so that it is relatively simple to separate the harmonic components.

An experimental determination [6] of frequency multiplier performance as a function of frequency has been investigated with a GaAs MESFET by Fujitsu (W_g = 600 μm, L_g = 1 μm) using a simple quarter wavelength open stub as a filter and input and output matching circuits.

The results of Figure 4.16 were obtained for an input power of +4 dBm, and adjusting both the tuning and bias voltages at each frequency point. One can conclude from this experimental result that as the operating frequency increases, the multiplication gain decreases and follows the same trend

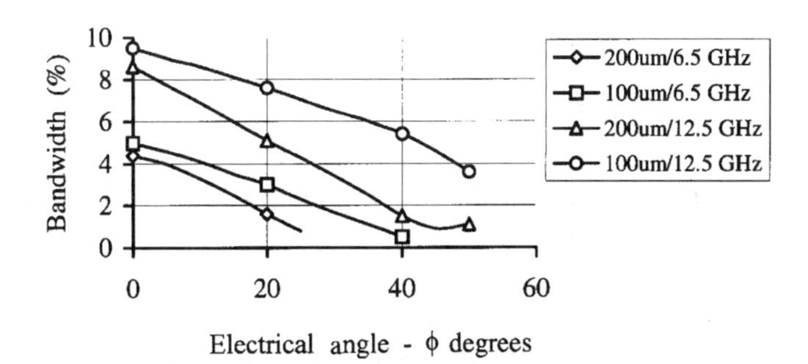

Figure 4.15 Bandwidth as a function of electrical length [4]. (*Copyright 1981 by IEEE.*)

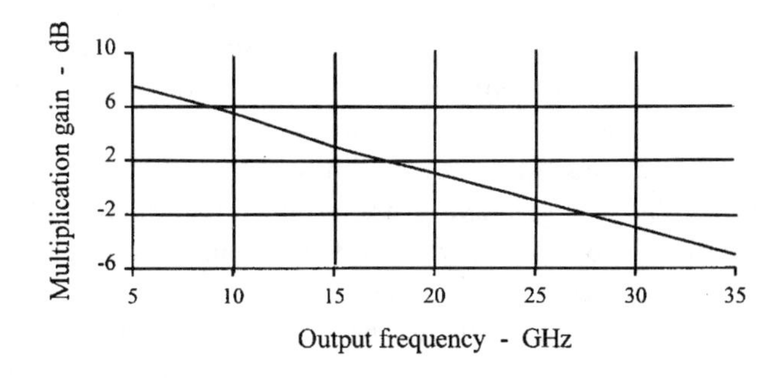

Figure 4.16 Performance of a MESFET frequency doubler [6].

presented by the maximum amplifier gain, either the stable gain defined by the ratio $|s_{21}/s_{12}|^2$, or the maximum available gain defined by $|s_{21}/s_{12}|^2[k - (k^2 - 1)^{1/2}]$.

Therefore, to design a wideband frequency multiplier, the matching circuit has to be optimal at the high end of the band, and mismatch the device impedance at the low end of the band in order to obtain a constant multiplication gain within the band. This problem is similar to the one found in wideband microwave amplifier design. A simple expression [7] has been derived to estimate what is the amount of power that has to be reflected from the matching network to obtain that objective,

$$|\Gamma|^2 = 1 - K_1(1 - \Delta\omega/\omega_H)^2 \tag{4.34}$$

where,

Γ = reflection coefficient from the matching network

K_1 = 1 for perfect matching at the high end of the band, otherwise it is ≤ 1

$\Delta\omega$ = frequency band

ω_H = high end of the band

This equation was derived based on the natural roll-off of available gain of FETs with increasing frequency, and the mismatch loss definition from a reactive network.

To conclude this chapter, it is interesting to include the experimental results reported [6] on the investigation of FM noise characteristics of two frequency doublers in cascade, showed the predicted effect of multiplication due to frequency deviation, expressed by 20log(N) where $N = 4$, with no excess noise degradation from the multiplier. However, in a similar experiment carried out by the author using a single device (measuring $250 \times 0.25\ \mu m^2$) as a frequency quadrupler, more than 5 dB excess phase noise was observed above the N^2 law, when the device was driven deep into saturation. This is a consequence of nonlinear AM to PM conversion that is observed in the multiplier gain and phase response as a function of drive level. The complete simulation of this effect is depicted in the plot of Figure 4.17.

Notice an increase in phase proportional to the input power until the output power starts to saturate. Increasing the power even further, the output power remains constant, but the phase response loses correlation with the input power. Considering this effect is a result of a large-signal applied to the gate-

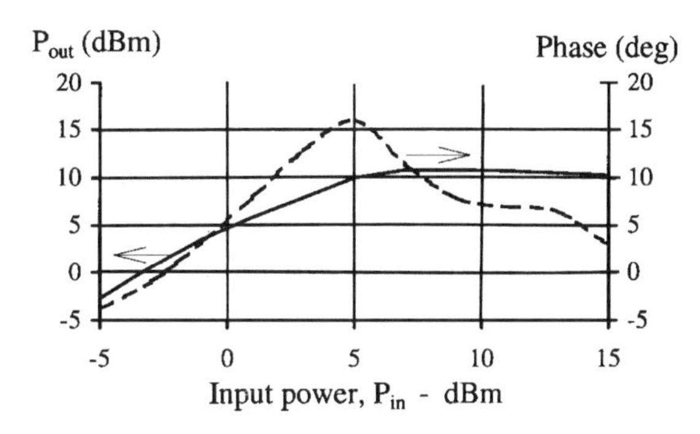

Figure 4.17 Power and phase versus input power.

source and gate-drain diodes, with a significative modulation of its capacitance, they result into excessive AM to PM conversion in the multiplier and therefore into phase noise degradation.

References

[1] Camargo, E., and F. S. Correra, "A High Gain GaAs MESFET Frequency Multiplier," IEEE MTT-S 1987 International Microwave Symposium, Las Vegas, June 1987, pp. 177–180.

[2] "S parameters: Circuit Analysis and Design," Hewlett-Packard Application Note 95, September 1968.

[3] Rauscher, C., "High Frequency Doubler Operation GaAs Field Effect Transistors," *IEEE Transactions on Microwave Theory and Techniques,* Vol. 31, No. 8, June 1983, pp. 462–473.

[4] Hirota, T., and H. Ogawa, "Uniplanar Monolithic Frequency Doublers," *IEEE Transactions on Microwave Theory and Techniques,* Vol. 37, No. 8, August 1989, pp. 1249–1254.

[5] Magnusson, P.C., *Transmission Lines and Wave Propagation,* Boston: Allyn and Bacon, Inc., 1965.

[6] Shima, T., T. Takano, T. Katoh, H. Sugawara, and H. Komizo, "Circuit Design and FM Noise Characterization of 20/30 GHz GaAs MESFET Multiplier Chains," European Microwave Conference, 1983, pp. 252–257.

[7] Ladbrooke, P. H., "MMIC Design: GaAs FETs and HEMTs," Norwood, MA: Artech House, 1989, pp. 264–267.

5

Design Strategies for High Frequency Multipliers

The initial phase in multiplier design is determining the feasibility of a desired performance in terms of multiplication gain and output power. If the desirable characteristics are achievable, what are the characteristics required from the active and passive elements to achieve that performance? This objective is a function of the device's nonlinear characteristics which is dependent on frequency of operation, level of input drive, bias conditions and terminations at the fundamental frequency, and the most important harmonics. The performance of a device at the fundamental frequency is the first factor to be taken into account when choosing the appropriate device. It can be evaluated from the transconductance, g_m, the transition frequency, f_T, defined as the frequency where the current gain is 1, and the maximum frequency of oscillation, f_{max}. The first parameter has a direct impact in the device's power performance and multiplication gain, and the last two parameters, determined from (5.1) and (5.2), give an insight into the importance of the parasitic elements associated with a given device and of how well it will fit into a given design.

$$f_T = \frac{g_m}{2\pi C_{gs}} \tag{5.1}$$

$$f_{max} = \frac{f_T}{2}\sqrt{R_{ds}/(R_g + R_i)} \tag{5.2}$$

The low-frequency nonlinear analysis developed in Chapter 3 introduced the basic nonlinear mechanisms within a device and how they can be used as a power amplifier or as a frequency multiplier. For instance, it was shown that from I_{DSS}, V_P and BV_{DS} for a particular device, one can estimate what power can be obtained for a given multiplication order. The same conclusions are applied to microwave frequencies, although at these higher frequencies the interaction of nonlinear elements with the external circuit is substantially different due to parasitics and feedback effects. Consequently, the performance of the nonlinear elements and the terminating impedances at the fundamental frequency and harmonics becomes inter-related. This effect precludes the development of an analytic solution to calculate the terminating impedances. Therefore, iterative solutions have to be used by one of the design procedures that is described in this chapter.

Once the device is selected, the task is to obtain the passive network that interacts with the device to provide the required performance. In this step, one has to model input and output impedances for designing the matching networks, a procedure similar to what is done in amplifier bilateral design. The differences in frequency multipliers are the constraints introduced by the load for enhancing harmonics, and the multiple number of frequencies to be considered in the design. The synthesis of the matching network is carried out by standard techniques long applied to amplifier design.

The objective of this chapter is to present the common strategies followed in determining the parameters required for the design of frequency multipliers for narrow band, wideband, power output, etc.

The Linearization Approach

Linearization is obtained by replacing the nonlinear elements of the equivalent circuit by fixed elements. Therefore,conventional linear analysis can be applied to design the matching networks. The final circuit can then be optimized by a full nonlinear analysis.

Direct Nonlinear Synthesis

In this method, the ideal terminations are applied to a simplified device model containing only intrinsic elements. Then, the resulting voltages and currents are transferred to the external terminals by adding the device parasitics, thus providing the optimum harmonic impedances to be connected to the device.

Computer Optimization

A computer optimization approach is introduced, where the harmonic power in a MESFET is described in terms of harmonic voltages and currents. Expressing the objective function in terms of desired output power at the harmonic frequency, for a given input at the fundamental frequency, optimization techniques are used to determine the set of voltages and currents that meet the desired requirements.

Harmonic Load Pull

This approach employs no device model and is essentially an experimental process. The device is inserted into a circuit that has the input tuned at the fundamental frequency and the output tuned for best performance at the desired output harmonic. The device is then removed and the networks are measured in the network analyzer, resulting in the desired impedances to be applied to the device. Then, conventional linear analysis is used to synthesize the matching networks.

Each of these methods should result in a high performance frequency multiplier even if the solution for a given problem provided by each one are not the same, since in nonlinear theory there might be more than one solution to a given problem [1]. The first two methods use simplified nonlinear techniques and do not require sophisticated nonlinear software. The third technique is based on full nonlinear optimization techniques requiring a dedicated software, and the last method relies on experimental techniques and specialized test equipment.

5.1 Linearization Approach

The linearization approach [2] consists in replacing the nonlinear elements from the equivalent model by fixed equivalent values at each harmonic frequency, providing the same harmonic voltages and currents as in the original circuit. Therefore, the multiplier circuit may be split into a fundamental frequency component and an output harmonic circuit, as depicted in Figure 5.1. This process can be applied to all harmonics of interest, but for the sake of circuit simplicity, it is applied here to only two circuits. Part (a) of the figure represents the fundamental frequency equivalent circuit, where the nonlinear elements, C_{GS} and G_M, were replaced by equivalent values C_{GS0} and G_{M0}. The generator impedance, Z_m, matches the input circuit at the fundamental frequency so that the gate voltage V_c is maximum. The drain circuit contains the fundamental frequency current, and the output reactance is adjusted to meet the optimum conditions for multiplication.

Part (b) of the figure represents the equivalent circuit at the output harmonic. The "link" between fundamental frequency and output harmonic is achieved by the control capacitor, C_{gs}. The multiplication transconductance, $G_{Mn} = I_{dn}/V_c$, gives the magnitude of output harmonic current which is proportional to the control voltage V_c. The gate circuit is terminated by impedance

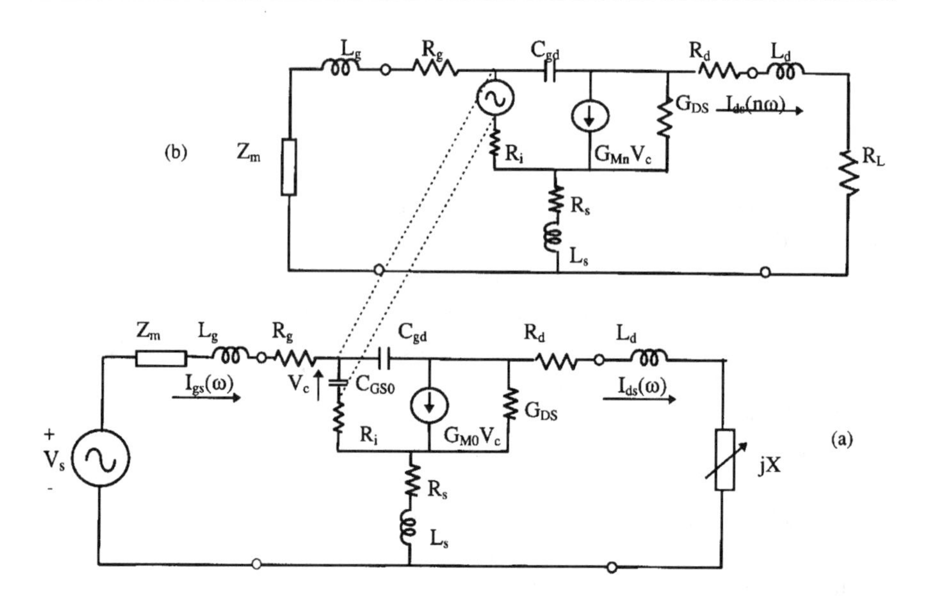

Figure 5.1 Fundamental equivalent circuit (a); and output harmonic equivalent circuit (b).

Z_m used to match the fundamental frequency impedance, and the drain circuit contains the load where the output power will be delivered. The main problem with this technique is determining equivalent values, since nonlinear elements are dependent on their terminal voltages, which in turn are dependent on the termination connected to them. In the case of a frequency multiplier, this task is greatly simplified due to restrictions on the drain termination at the fundamental frequency. A more appropriate name for this technique would be *quasi-linearization* [1], since this process is valid only for a specific bias, drive level, and particular drain terminations. Modification of any of these parameters requires a new analysis. The linearized model provides S-parameters that can be used in the design of matching networks, to analyze the circuit stability and to determine the multiplication power gain.

The linearization steps are described in Figure 5.2. The first step in this process is the determination of the Fourier coefficients for the nonlinear elements within one cycle of the applied signal.

The arrangement of Figure 4.8 is used to determine the equivalent input impedance of a MESFET model, carrying out simulations at low-frequency where the parasitics can be neglected and the generator voltage is directly applied to the control capacitor. This procedure is valid if caution is taken to maintain the low frequency amplitude, V_c, the same as at high frequency. That process is also valid in determining the Fourier coefficients of a nonlinear MESFET model for a given signal amplitude. The linearized transconductance

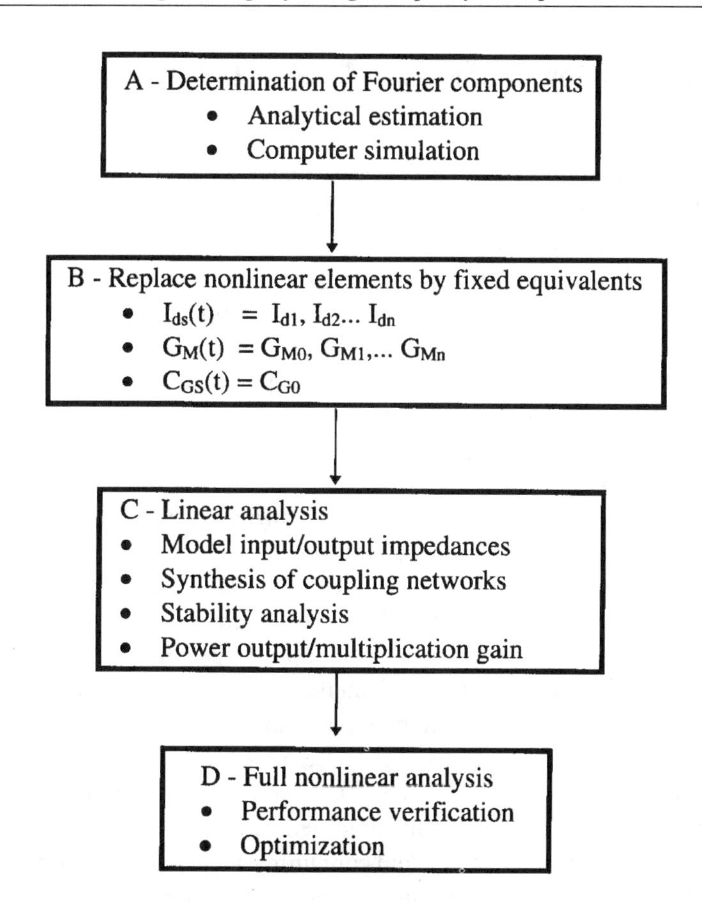

Figure 5.2 Linearization procedure.

parameters are determined by simply relating the drain current components to the gate voltage for a shorted circuit drain:

$$G_{M0} = I_{d1} / V_c = \text{Large-signal transconductance}$$

$$G_{M1} = I_{d2} / V_c = \text{First order multiplication transconductance}$$

$$G_{M2} = I_{d3} / V_c = \text{Second order multiplication transconductance}$$

The determination of output conductance is more complex, due to its dependency on output impedance. In the case of a class B multiplier, the low drain impedance shunts this element and it can be considered constant at the output harmonic. In class A or I_{DSS} multiplier, its value has to be determined iteratively, searching for its dependency as a function of load. The details of the linearization approach are described in the following example.

5.2 Linearization Techniques Applied to a Frequency Doubler

The objective is to design a narrow band frequency multiplier to double 5 to 10 GHz for maximum output power. Therefore, the gate voltage is expected to swing from pinch-off to zero volts, the peak drain current is equal to I_{DSS} and the peak drain voltage is equal to $(V_{DD} - V_k)$. The multiplier performance parameters are the multiplication gain, stability of operation, and input/output match. Matching the output at the second harmonic provides additional multiplication gain and power. However, the output match parameter usually is not considered a performance parameter in system design, since its effect on overall performance is of second order of importance. Besides, its measurement requires a complex harmonic load pull set-up [3]. Therefore, only the input match will be considered as a multiplier performance parameter.

The design starts with the definition of the operating point. The low frequency model suggests biasing the device in class C with a conduction angle of 120 degrees for best efficiency. However, at this bias the FET channel is completely pinched-off and the resulting input impedance is more difficult to match to the generator at the fundamental frequency of operation. Besides, large reverse voltages on the gate may result in an avalanche of the drain-gate diode. Therefore, let us consider the device biased in class AB, near pinch-off, with a low quiescent drain current ($\approx 0.1\ I_{DSS}$) which is a tradeoff condition between input matching and harmonic generation. In these conditions, the output current is a half-wave sinusoid containing a low number of odd harmonics, so that the fundamental frequency and second harmonic frequencies are the ones considered in the design. In this example, $V_{DD} = 3.5$ Volt, $V_P = -1.5$ Volts and $I_{DSS} = 60$ mA.

The second step is to determine the linearization parameters. They can be estimated using the dc model of Chapter 3 or they can be simulated at low-frequency employing Pspice [3]. The results are in Table 5.1. The assumption is that only the capacitance and transconductance are the important nonlinear parameters, and the linearization takes place by replacing them by the values described in the table. This approximation gives good results, but better accuracy

Table 5.1
Linearization Parameters

Freq(GHz)	G_{M0}(mS)	G_{M1}(mS)	C_{GS0}(pF)
DC	20	10.0	0.3
0.1	17	8.5	0.28

is obtained by iterative procedures, where the initial values are corrected after the first iteration. It is important to monitor the magnitude of the control voltage all over the procedure, to guarantee the validity of the linearization.

The third step is the definition of the input and output impedances at the fundamental frequency and second harmonic, used to design the matching networks. It is in this step that the linearization is more useful. The procedure starts with definition of the output termination then the input impedance matching at the fundamental frequency and at the harmonics.

5.2.1 Fundamental Frequency

5.2.1.1 Harmonic Load

The drain termination at the fundamental frequency is most important for any type of frequency multiplier, since it controls the output wave shape and has a direct impact on several multiplier parameters. From low-frequency considerations, a low impedance is required at the fundamental frequency and third harmonic, and the second is matched to the optimum load. An example of such a load built with transmission-line elements is represented in Figure 5.3.

The association of open-circuit stubs, 90 degrees at the fundamental frequency, separated by a series of lines of the same length apply a short circuit on the main line at the fundamental and third harmonic. If the impedance of all lines are equal to the reference, 50 ohms, then low losses are introduced at the second harmonic. The purpose of the series line of electrical angle, ϕ, inserted between the drain and the harmonic load is to series resonate the device output reactance at the fundamental frequency, therefore applying a low impedance at the internal drain terminal. To verify the effect of load reactance at the nonlinear current source terminals, the internal voltage $(V_{di} - V_{si})$ was calculated using a linear software package, Touchstone® [4],

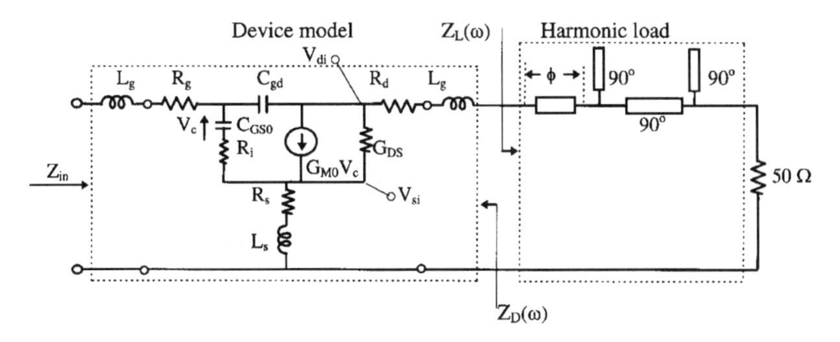

Figure 5.3 Harmonic load for a frequency doubler.

as a function of the series line phase. The transconductance was linearized employing the dc value described in Table 5.1 and the result is plotted in Figure 5.4.

Minimum voltage is obtained at ϕ = 162 degrees corresponding to reactance equal to $X_L = jZ_0\tan(162°) = -j16.2\ \Omega$ at the fundamental frequency of 5 GHz. The minimum internal voltage is also obtained with a series capacitor in place of the series line. A capacitance of 1.96 pF would have the same effect. The parallel resonance with C_{gs} and C_{gd} is obtained at $\phi = 72°$, or $X_L = 153.8\ \Omega$, when maximum voltage is observed. The optimum drain termination at the fundamental frequency is given by:

$$Z_L(\omega) = 0 + jX_L \tag{5.3a}$$

$$Z_D(\omega) = R_{DS} + jX_D \tag{5.3b}$$

$$X_L = -X_D \tag{5.3c}$$

5.2.1.2 Input Impedance

An approximate expression for the input impedance when the drain is shorted to the ground is given by (4.18), providing a good initial point for impedance calculations, and is reproduced below,

$$Z_{in} = R_g + R_s + G_{M0}L_s/C_{GS0} + 1/(j\omega C_{GS0}) \tag{5.4}$$

Applying the numeric values of Table 2.2 to obtain the input impedance at the fundamental frequency using the values of Table 5.1 results in

$$Z_{in}(\omega) = 15.0 - j106.0 \text{ ohms, using dc values}$$

$$Z_{in}(\omega) = 16.4 - j80.0 \text{ ohms, using 100 MHz values}$$

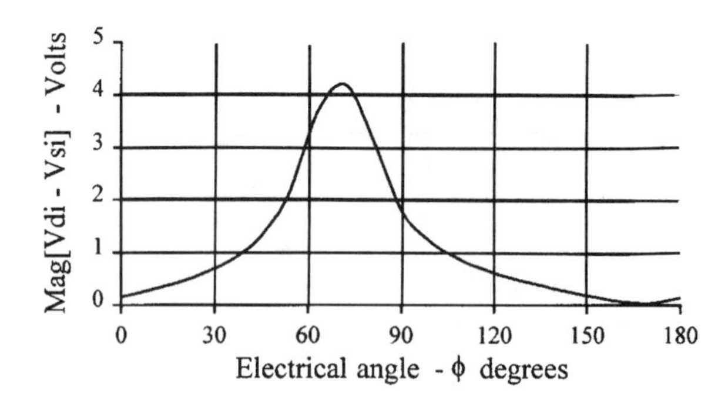

Figure 5.4 Internal drain: source voltage magnitude versus line angle.

The drain load impedance has a limited bandwidth due to the open-circuited stub tuned characteristic, so that the input impedance will deviate from these values as a function of frequency. The effect of harmonic load on input impedance has already been qualitatively determined in the previous chapter. But in order to quantify this effect for this example, the input impedance was simulated as a function of phase of the short-circuit connected to the drain of a linearized MESFET. The results are shown in Figure 5.5.

Observe that the reactance is always capacitive changing from 100 to 200 ohms when the short-circuit phase changes from 0 to 180 degrees. The resistive part presents a more drastic change, starting with a positive value when the short-circuit is close to the drain terminal, $\phi = 0°$, changing to a negative value when $\phi = 18°$, and turning back to positive when $\phi = 72°$. Maximum negative value peaks at $\phi = 54°$. The negative resistance at certain frequencies is created by the feedback introduced by the capacitive divider C_{gd}, C_{gs}, which is in parallel with an inductive load.

A lowpass matching network indicated in Figure 5.6 was employed to adapt the input impedance at the fundamental frequency to the generator impedance. These values were determined by the equations described in Appendix B, where other network options are described for matching complex impedances. The advantage in using a lowpass network is the high rejection of harmonics at the generator port. The input power can be estimated assuming the total power is absorbed by the effective input resistance. In the particular case of maximum peak gate voltage, (i.e., $V_c = V_P$), the input power is approximated by

$$P_{\text{in}} = 0.5(\omega C_{GS0} V_P)^2 \, (R_g + R_s + G_{M0} L_s / C_{GS0}) = 1.55 \text{ mW } (1.9 \text{ dBm}) \tag{5.5}$$

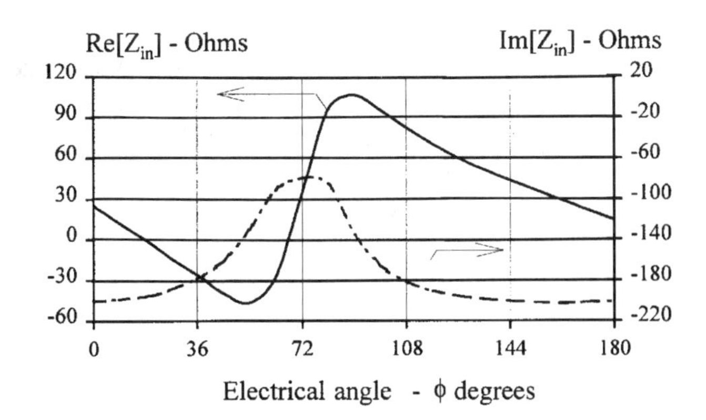

Figure 5.5 Simulated input impedance as a function of short circuit phase.

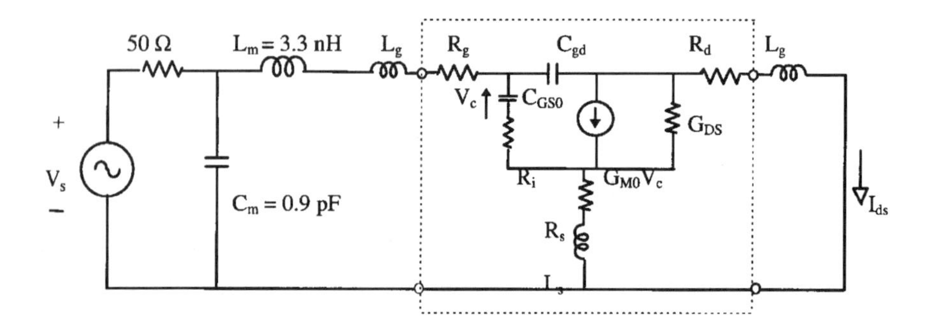

Figure 5.6 Equivalent circuit at the fundamental frequency.

5.2.2 Second Harmonic

The circuit represented in Figure 5.7 depicts the FET linearized for second harmonic frequency, where the magnitude of generator V_c is determined from fundamental frequency analysis and G_{M1} is determined from the ratio of second harmonic drain current to fundamental frequency gate voltage.

The fundamental frequency voltage generator represented in Figure 5.7, V_c, should be made as high as possible for best efficiency. Therefore, the gate impedance should ideally be a parallel resonator at the second harmonic. In this case, the low pass L matching network designed for the fundamental frequency, presents a high impedance at the second harmonic, meeting the desired second harmonic termination at the gate.

The output impedance at the second harmonic for a short-circuited gate is obtained by simplifying the circuit of Figure 5.7 to the one represented in Figure 4.3, resulting in

$$Z_{\text{outsc}}(2\omega) = R_s + R_d + j\omega(L_s + L_d) + R_{DS}/[1 + j\omega R_{DS}(C_{GS} + C_{DS})] \tag{5.6}$$

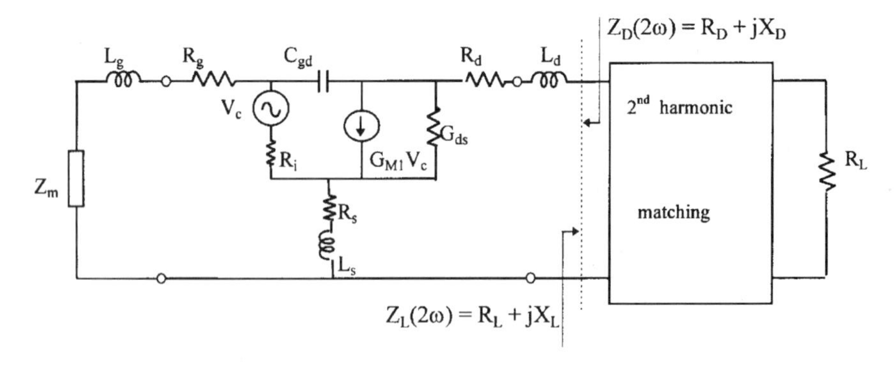

Figure 5.7 Equivalent circuit at the second harmonic.

In the case of an open-circuit connected to the gate, the capacitances C_{GS} and C_{GD} become in series and the feedback effects have to be accounted for. Thus, the output impedance becomes,

$$Z_{outoc}(2\omega) = R_s + R_d + j\omega(L_s + L_d) + 1/[G_{DS} + G_M C_{DG}/C_{GS} + j\omega C_{DS})] \quad (5.7)$$

The matching networks are selected to series resonate the device output reactance and to provide a real part that gives the maximum output power. The matching conditions are obtained by making $Z_L(2\omega) = Z_D(2\omega)$.*

5.2.3 Simulation Results

The circuit stability was carried out looking at the small-signal input impedance of the multiplier circuit with a shorted drain, and checked for a negative resistance. The results are in Figure 5.8 which shows that the real part is positive within the band.

The input impedance obtained from Pspice using a full nonlinear model is also shown where it is observed they follow each other over the band. One also has to investigate the drain impedance with the gate terminated with the input matching network. That impedance is in the same plot and is positive within the bandwidth but shows a tendency to become negative at lower frequencies. At lower frequencies, resistive terminations can be added to the drain and gate bias, absorbing negative resistances and stabilizing the circuit.

The simulation of output impedance using the linearized model and the parameters from the reference device, provided, $Z_D(2\omega) = 63 - j\,60$ ohms. Therefore, a 63 ohms resistor in series with a small inductance ($L_m = 0.9$ nH) provides an optimum load for this circuit. The design challenge is to integrate a transformer from 50 ohms to R_L, within the harmonic load

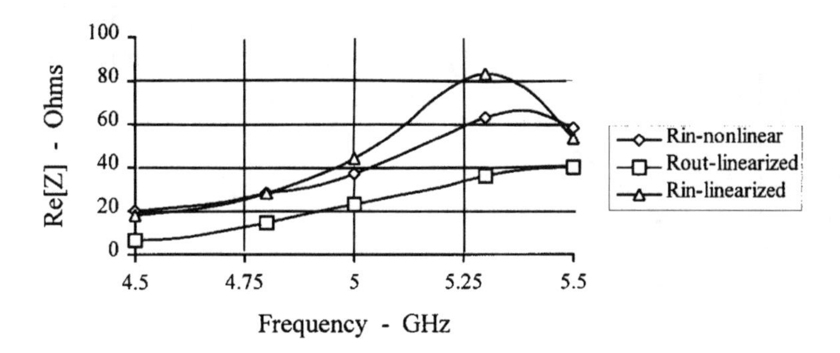

Figure 5.8 Real part of input and output impedance at 5 GHz.

circuitry. Even though it is possible to associate two circuits in cascade for that purpose, losses are increased, the circuit presents a more tuned characteristic, and the advantage of optimum load is lost. Thus, in this example, a 50-ohms termination with a small series inductance is used as the optimum load. The series transmission-line T_{L5} is used to adjust the phase of the second harmonic circuit so that there is a integer multiple of 180 degrees between the load and the drain. It is possible to estimate the maximum power on the load driven by the second harmonic current source I_{d2}, employing (3.33) and assuming a lossless matching circuit. Using the values for the reference device, one obtains,

$$P_{out} = \frac{Re(I_{d2}V_{d2}^*)}{2} = \frac{R_L I_p^2}{2} = \frac{R_L}{2}\left[\frac{2I_{DSS}}{3\pi}\right]^2 = 4.05 \text{ mW (6.1 dBm)} \quad (5.8)$$

The simulation of multiplication gain performance was done in two steps using a linear simulator. In the first step, the fundamental circuit is simulated to obtain the magnitude of the control voltage, $|V_c|$, over frequency as a function of applied power. In the second step, the second harmonic circuit is simulated using the value of control voltage determined in the first step to calculate the power at the load. The results are in Figure 5.9, which compares the multiplication gain to the full nonlinear model using Pspice.

Again, a reasonable agreement is observed between the linearized simulation compared to the full nonlinear simulation employing Pspice. The simulation of input and output return loss were applied to the linearized model. The results of Figure 5.10 show that the output circuit resembles that of a tuned circuit, due to long transmission-line between the matching and the output circuit. Note that output return loss is at the second harmonic, so that the horizontal scale has to be doubled to obtain the correct frequency.

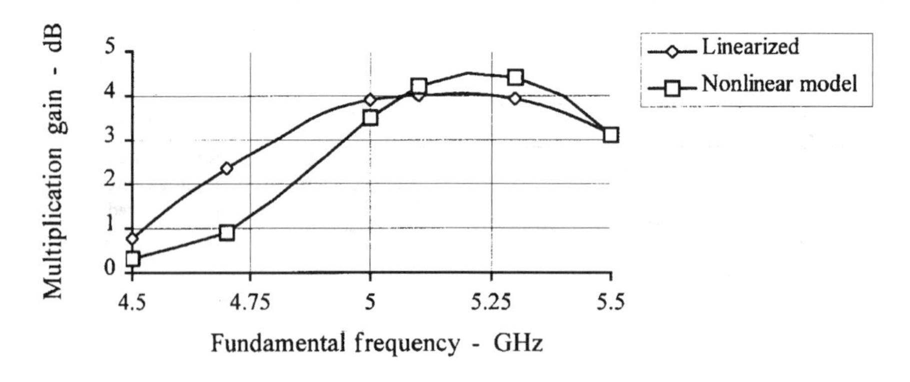

Figure 5.9 Multiplication gain for linearized and full nonlinear model.

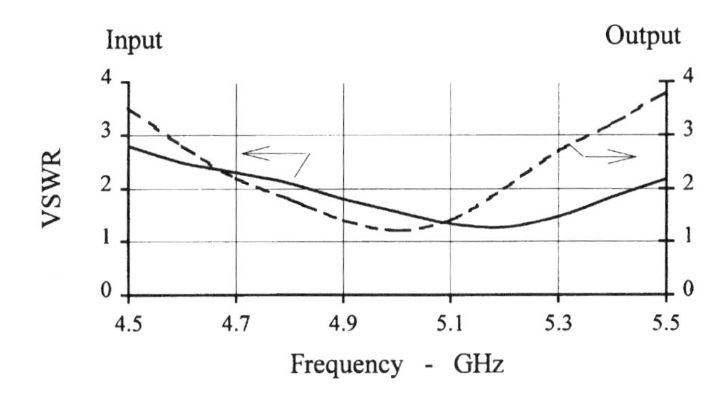

Figure 5.10 Input and output VSWR for linearized model.

The complete circuit is shown in Figure 5.11, including biasing circuits. The input circuit contains a capacitor, C_{bl1}, to block the dc current, the matching elements, L_m, C_m, and the bias network. The isolation between the bias and the inband signals is provided by the transmission line $T_{L1}(Z_0 = 70\ \Omega/\phi = 90°)$ which is a quarter-wavelength long at the fundamental frequency. The bypass capacitor C_{bp1} applies a low impedance to T_{L1} at the fundamental frequency, and due to the transformer action, it is an open circuit at the gate. A capacitor in the order of 20 pF is a reasonable choice. The bias is completed by a 50-ohms resistor and another high value capacitor, C_{bp2}. The objective is to create a very low pass frequency response with a lossy circuit that can absorb eventual negative resistance generated by the device.

The output circuit contains a transmission-line phase shifter, a bandstop filter similar to the one employed in Figure 3.3, a dc blocking capacitor, C_{bl2}, presenting a low impedance at the second harmonic and the bias circuitry.

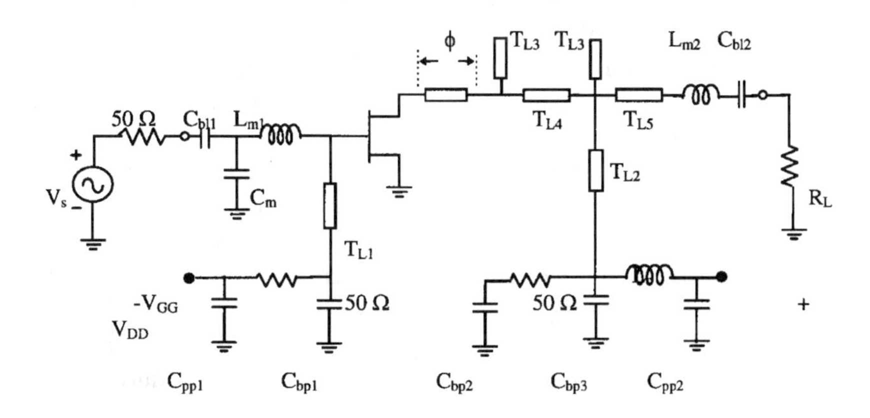

Figure 5.11 Schematic of complete frequency doubler.

The bias line, T_{L2}, is of the same impedance as line T_{L1} but it is half its length. Observe the position where the bias is applied, after the band reject filter, in order to avoid connecting the bias close to the drain and create an interference with the generated harmonics. In this case, it is important to avoid dc current from flowing through the 50 ohms resistor to avoid unnecessary dc power dissipation. Thus, a bias inductance, L_b, is employed to provide additional isolation from the RF circuit to the external power supply. A final note in this topology is on the bandstop filter connected to the drain which has no means to prevent low frequencies from being amplified by the device. Therefore, it is recommended to add a bandpass filter with a minimum number of resonators in cascade with the bandstop filter for this purpose.

This exercise demonstrates the usefulness of the low frequency analysis developed in Chapter 3, and the simplicity of linearization techniques applied to a frequency doubler design. The efficacy of such an approach has been demonstrated by the good agreement obtained between this approach and the one obtained by a sophisticated tool such as Pspice.

5.3 Direct Nonlinear Synthesis

This method is based in a previous work [5] developed for oscillator design, and can be extended to other nonlinear circuits such as power amplifiers, frequency multipliers, harmonic oscillators, etc. The direct synthesis technique relies on the determination of optimum voltage and current conditions, close to the device nonlinear elements, that will enable the desired performance. Therefore the simplest FET model, representing the device embedded into the GaAs, has to be considered. Once those conditions are defined, they are transferred to the external terminals taking into account the parasitics. The four steps involved in this process are described in Figure 5.12.

The simplest FET for multiplier design is represented in Figure 5.13. It includes the nonlinear current source, the gate-to-source $R_i - C_{gs}$ circuit and the series parasitic resistances, R_s and R_d.

A simpler model where R_d and R_s are suppressed can be used in the design of power amplifier and oscillators. In the case of frequency multipliers, they are required to avoid solution with negative resistance when applying a short circuit at the drain. The solution to which set of voltages and currents provides the required performance can be found by computer optimization, which runs efficiently due to the simplicity of the circuit. An alternative simplified process using the conclusions of Chapter 3, is introduced next.

The optimum drain harmonic load for a frequency multiplier is provided by a load that presents the following impedances:

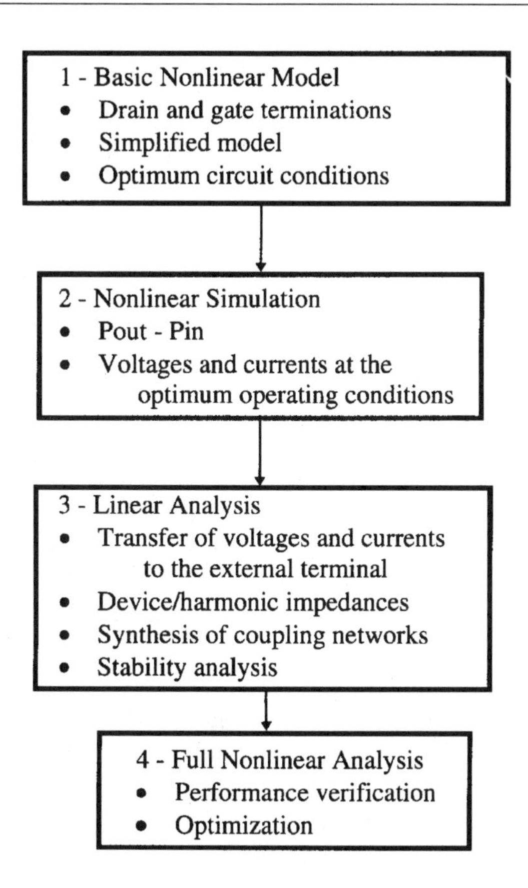

Figure 5.12 Design steps for direct synthesis.

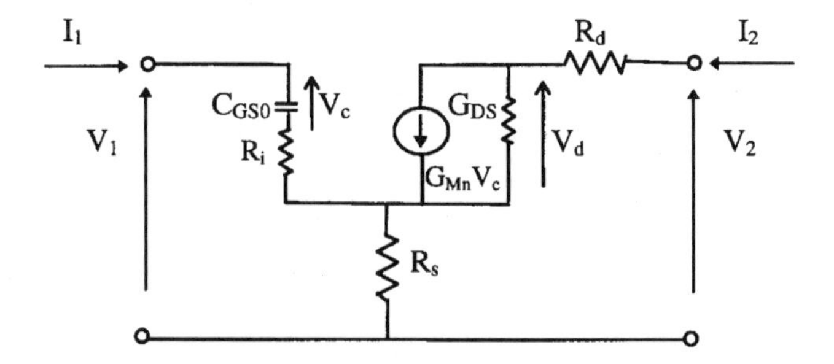

Figure 5.13 Simplified FET model.

$Z_L(\omega_0)$ = short circuit for distorted drain current
open circuit for distorted drain voltage

$Z_L(n\omega_0) = (V_{DD} - V_k)/I_{dn}$, distorted drain current
$= 2V_{dn}/I_F$, distorted drain voltage

$Z_L(m\omega_0)$ = all other harmonics are reactively terminated with a proper phase for best performance, $m \neq n$

The input impedance is matched to the generator impedance at the fundamental frequency by a simple R–L network and the generator voltage is such that $V_c = V_P$. A nonlinear simulation is carried out with a few iterations to correct the initial value for the load impedance and to obtain a good input match over the band. When the multiplier performance, in terms of output power, multiplication gain, and impedance match is satisfactory at a given iteration, one has determined the optimum set of voltages and currents in the simplified circuit at the fundamental frequency and harmonic frequencies.

The set of voltages and currents (I_1, V_1, I_2, V_2) are then transferred to the external terminals, by adding the parasitic elements, as shown in Figure 5.14. A simple linear software program can perform this task. It is also used to determine the relations between voltages and currents, (therefore impedances) at the output terminals at the fundamental frequency and harmonic frequencies. The effect of feedback elements are intrinsically accounted for in this process.

The input and output power are obtained by applying (3.33), also providing the multiplication gain. This design procedure is the most efficient compared to the other approaches, since the nonlinear simulation is carried out for a simple circuit with a very fast convergence. All remaining calculations are carried out with linear simulators.

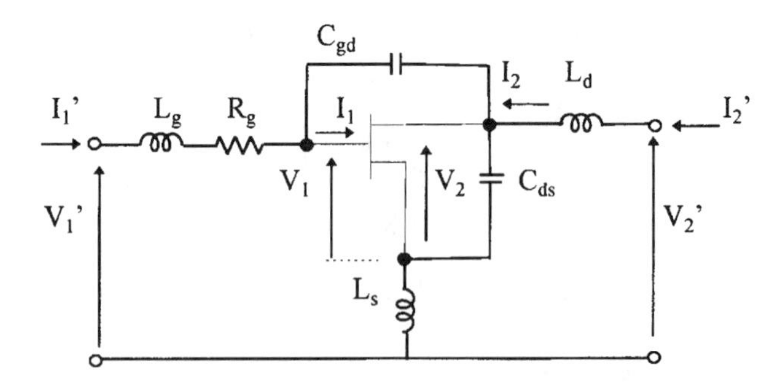

Figure 5.14 Transference of internal to external variables.

5.4 Direct Synthesis Applied to a Frequency Tripler

The objective is to design a narrow band frequency tripler from 5 ± .1 GHz to 15 ± .3 GHz for maximum gain and output power, corresponding to the condition of maximum efficiency.

There are two ways of generating a current waveform rich in third harmonic: by biasing the device in class C with a conduction angle of 80 degrees and by biasing the device at class A and overdriving the device until there is clipping due to pinch-off on the negative swing and due to gate conduction on the positive swing. The class C approach presents the same restrictions considered in the previous example and is not the adequate bias choice. The second option is to use class A bias with a high gate resistor to limit the gate dc conduction and assure reliable operation. A third option consists of generating a voltage waveform rich in odd harmonics. This option is obtained by biasing the device in class A and connecting a high impedance to the drain at the fundamental frequency to generate a distorted voltage waveform. Therefore, the drain waveform is compressed at low-drain voltage by the device resistive region while at high voltages it is compressed by the nonlinear output conductance. In order to compare the efficiency of each one of these approaches, a low-frequency nonlinear simulation was carried out to investigate the generation of third harmonic current. The obtained results are in Table 5.1.

Obviously, the voltage distortion approach provides higher odd harmonic current and was selected for this exercise. The device is biased at $V_{GG} = -1.0$ V, $V_{DD} = 3.5$ V. The dynamic gate voltage is equal to 1.5 V in any condition, enough for sweeping the gate from pinch-off up to +0.5 V.

5.4.1 Harmonic Load

The distorted voltage waveform has to be as symmetrical as possible, so that even harmonics are minimized. Ideally, the harmonic load has to present a

Table 5.1
Third Harmonic Current Generation

Mode of Operation	I_{d3} – mA
Current, class C	4.2
Current, class A saturated	4.6
Voltage, class A	12

short-circuit for the even harmonics and an open-circuit at the odd harmonics, except the output harmonic, which is terminated with a load for maximum output power. A bandpass filter in series with a phase-shifter can approximately present such terminations. The phase-shifter adjusts the filter impedance to a high value at the fundamental frequency and is transparent to the third harmonic. The resulting impedance at the second harmonic will be dependent on the filter impedance. In this example the transmission-line phase-shifter is adjusted to present a high impedance to the drain at the fundamental frequency.

5.4.2 Input Impedance

The arrangement for measuring the input impedance is displayed in Figure 5.15 where the simplified model of Figure 5.13 is used for the device, and the harmonic load is connected to the drain. The impedance is determined by applying a voltage equal to 1.5 V at the control capacitor terminals and reading the resulting gate current. That step was carried out by nonlinear analysis with Pspice, and the following impedance was obtained at the fundamental frequency of 5 GHz,

$$Z_{\text{in}}(\omega) = V_{si}/I_g = 7.6 - j89.3 \text{ ohms}$$

The next step is to transform the circuit of Figure 5.15 to the circuit of Figure 5.16, whose generator internal impedance is $Z_s = Z_{\text{in}}^*$ and its amplitude

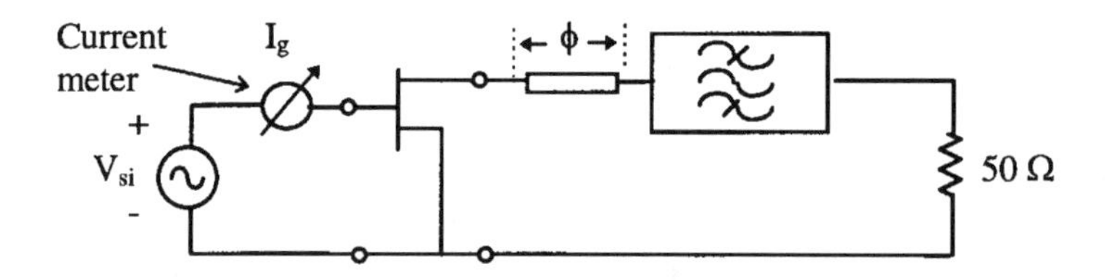

Figure 5.15 Determination of frequency tripler input impedance.

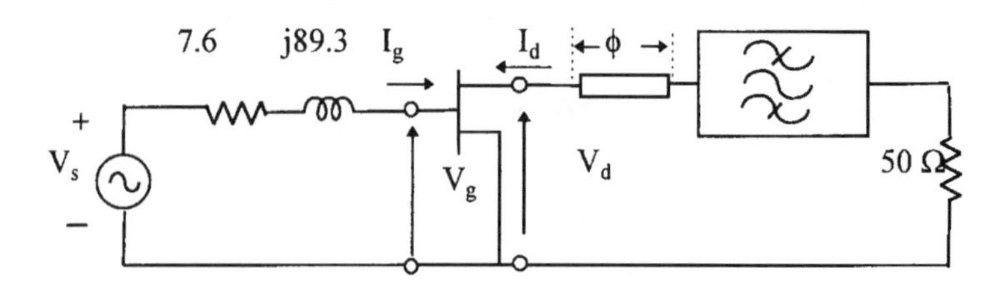

Figure 5.16 Simplified frequency tripler circuit.

is V_s. The magnitude of the voltage generator is equal to the voltage drop on the 7.6 ohms resistors, created by the gate current (I_g = 1.2 + j13.9 mA). Thus,

$$|V_s| = 2R_{\text{in}}\,|I_g| = 2(7.6)|1.2 + j13.9| = 0.22 \text{ V} \tag{5.9}$$

Assuming this is the optimum termination for the simplified model, a simulation was carried out to find out which are the voltages and currents phasors at the gate and drain of the device. The components are related on Table 5.2, where the first number is the phasor magnitude and the second is its phase.

The output power at the third harmonic and the power at the fundamental frequency can be obtained from this table by applying (3.33)

$$P_{\text{out}} = \frac{Re(I_{d2}V_{d2}^*)}{2} = \frac{Re[(13/-104.5)(0.64/75.5)^*]}{2} \tag{5.10}$$

$$P_{\text{out}} = 4.16 \text{ mW (6.2 dBm)}$$

The gate power is equal to +0.2 dBm, so that a multiplication gain of 6.0 dB is predicted by this method. The described procedure can be repeated for several gate peak voltages, so that multiplication gain curves as a function of input power at the fundamental frequency are obtained. It can also be repeated for several fundamental frequencies to obtain the multiplier's frequency response.

The procedure continues by transferring this set of voltages and currents to the external terminals, by applying the linear transformation [6] described in Appendix C. The transformation consists of adding the parasitic parallel elements as a "PI" network and series element as a "TEE" network. The former consist of adding the drain-to-gate (0.04 pF) and drain-to-source capacitance (0.1 pF), and the latter on adding series bonding inductances ($L_g = L_d$ = 0.2 nH; L_s = 0.15 nH), and gate resistance (R_g = 4.0 ohms). This is a very convenient process, easily able to account for any parasitics in the

Table 5.2
Voltages and Currents at the Device Terminals

Freq (GHz)	I_g (mA)	V_g (Volts)	I_d (mA)	V_d (Volts)
5.0	20.0/–2.7°	1.57/–87.1°	0.39/174.8°	3.62/83.9°
10.0	2.38/90.2°	0.375/–2.9°	15.0/–15.8°	0.03/–105.8°
15.0	0.15/–179°	0.037/88.6°	13.0/–104.5°	0.64/75.5°
20.0	0.035/–34°	0.011/–126°	2.7/–150.3°	0.06/–60.3°

chip assembling and also any package parasitics. In this example a chip device is under consideration so that the external voltages and currents at fundamental frequency and harmonics are related in Table 5.3.

The equivalent input and output impedances at each harmonic frequency are provided in Table 5.4.

The impedances shown are the ones required to obtain the desired internal voltages and currents. They are complex impedances at all frequencies. To create a network capable of presenting those impedances at each frequency is not practical, so that the network has to be designed in hierarchical steps.

1. *Fundamental frequency*: Terminate the drain with reactance that maximizes the voltage at the output conductance. A trade-off with circuit stability is required to avoid oscillations. Match the input impedance for maximum voltage at the control capacitor.
2. *Third harmonic*: Match the drain impedance for maximum output power.
3. *Second harmonic*: Use the bias network and the properties of transmission lines. For instance, a quarter-wavelength transmission line separates fundamental frequency and odd harmonics from even harmonics in a biasing circuit. Therefore, second harmonic at the drain can

Table 5.3
Voltages and Currents at the External Terminal

Freq (GHz)	I_g (mA)	V_g (Volts)	I_d (mA)	V_d (Volts)
5.0	26.4/–2.86°	1.36/–86.0°	18.3/–5.1°	3.54/84.0°
10.0	3.35/90.8°	0.36/19.1°	15.8/161.6°	0.29/74.6°
15.0	2.1/–18.0°	0.18/–18.0°	15.0/42.9°	0.58/28.2°
20.0	0.27/–133.1°	0.05/–69.1°	1.6/28.0°	0.14/–60.6°

Table 5.4
Harmonic Input/Output Impedances

Freq (GHz)	$Zin(\omega)$ ohms	$ZL(\omega)$ ohms
5.0	5.8 – j51.0	3.1 + j194.0
10.0	33.4 – j103	1.8 – j19.0
15.0	83.2 – j24.0	36.7 – j9.5
20.0	76.1 + j156.0	2.1 – j82.9

be adjusted independently of the fundamental frequency and third harmonic.

4. *Other harmonics*: The termination of other harmonics at the drain and gate are last in this list of priorities.

In this example only the principal impedances are matched, which modifys the initial assumptions for the internal voltages. The drain resistances at the harmonics related in Table 5.4, are small and can represent actual circuit losses. The gate impedances,on the other hand,are quite high and difficult to match simultaneously with a simple circuit.

The input network at the fundamental frequency is matched with a simple lowpass L-shaped network, applying the equations of Appendix B. The initial step is to transform the input impedance to 50 ohms by employing a series inductance L_m = 1.62 nH, and a parallel capacitance C_m = 1.75 pF. This network presents a high impedance at the second harmonic, so that it can be paralleled with another low impedance network to accomplish the second harmonic matching.

The output network contains a series line which acts as the phase-shifter and an edge-coupled bandpass filter. The phase-shifter has to resonate the equivalent FET output capacitance at the fundamental frequency, and its phase is calculated by using the reactance information from Table 5.4, (i.e., $\phi = \tan^{-1}(194/50) = 75.5$). Additional harmonic impedance matching still has to be done after the filter, which suggests building a filter for 37 ohms impedance instead of 50 ohms. Therefore, in this example, R_L = 37 ohms.

The gate bias filter contains a transmission-line filter, T_{L1} (70Ω/90°) and T_{L2} (30Ω/90°), that will block the fundamental frequency and is transparent at the second harmonic. Note that the quarter-wavelength bias filter becomes half-wave at the second harmonic, so that its terminating impedance is transferred to the gate terminal. The terminating impedance is composed of a lowpass circuit (L_{b1}=1.64 nH, L_{b2} = 0.24 nH C_{bp2} = 0.72 pF) which transforms 10 ohms to the second harmonic impedance = 33.4 – j103 ohms.

The drain bias filter, in this case, consists of lines T_{L3} (70Ω/30°) and T_{L4} (70Ω/30°) that block the third harmonic. An additional line, T_{L5} (70Ω/13°), is introduced to block the fundamental frequency. The electrical angles are still referenced to the fundamental frequency. Such a circuit is represented in the final topology depicted in Figure 5.17.

The transfer process can be applied one step further, to incorporate the input and output matching circuits, obtaining voltages and currents at the generator and load. The resulting input impedance is equal to Z_{in} = 46.7 – j0.6 ohms and the generator voltage is equal to 0.435 V. That

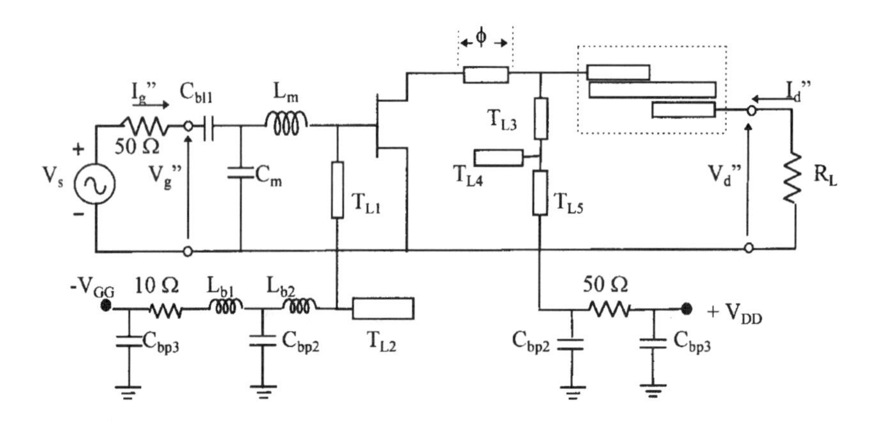

Figure 5.17 Schematic of complete frequency tripler.

corresponds to an input power of +3.0 dBm. The output power at the 37 ohms load is equal to +6.62 dBm. Comparing these values to the ones generated by the simplified circuit, we observed a reduction of 0.5 dB in the output power and an increase of 3 dBm in the fundamental frequency drive. This degradation is the result of feedback introduced by the parasitics.

A complete simulation of this circuit on Pspice is shown in Figures 5.18 and 5.19. The former shows the input impedance and provides insight into the stability of this circuit. It is observed that the circuit is matched around 5.0 GHz and the real part of impedance tends to negative values outside the operating bandwidth.

The multiplication gain is shown in the next plot, where a maximum multiplication gain of 7.0 dB at the center frequency can be observed. The harmonic drain resistances from Table 5.4 were suppressed from the matching circuit to obtain this result. The resulting operating bandwidth for a minimum input return loss of 10 dB is from 4.95 to 5.15 GHz. The output return loss

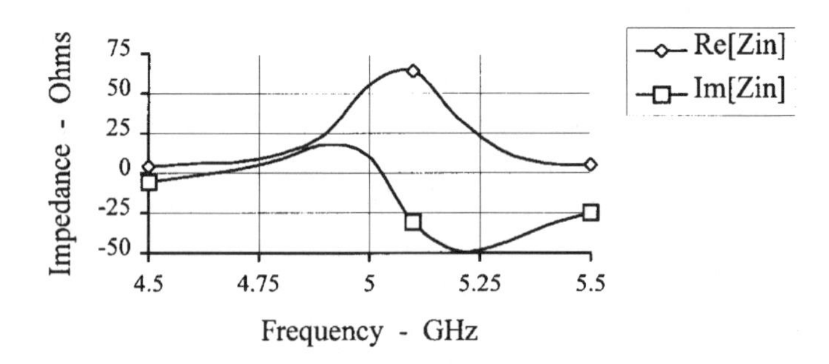

Figure 5.18 Input impedance as a function of frequency.

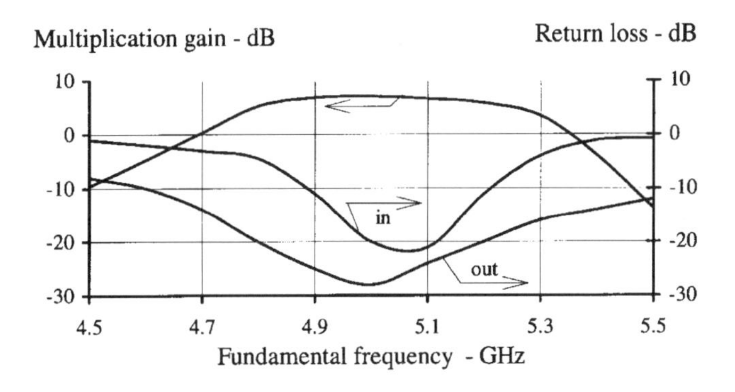

Figure 5.19 Multiplication gain and return loss versus frequency.

is in the same figure and was obtained by nonlinear simulation of Figure 5.16 for a few frequencies within the band, and by transferring the voltages and currents to the external terminals. The filter response was assumed to be ideal in this simulation.

5.5 Computer Optimization

The modern nonlinear simulators are capable of analyzing nonlinear networks e.g., Pspice and others such as LIBRA® [7], and MDS® [8] are also capable of optimizing them, employing methods similar to the ones established for optimizing linear networks. This process, however, is usually time consuming, may not lead to an optimum network, and may not converge if the circuit presents stability problems.

One option in the case of convergence problems is to introduce stabilizing resistors in strategic locations in order to make the circuit stable. Those points can be found by linear analysis. A nonlinear optimization is then applied to obtain the desired objectives. That approach will result in low multiplication efficiency due to the stabilizing resistors and a tradeoff will have to be made between stability and performance with resistor values as a parameter.

A more efficient computer optimization was recently introduced in the literature [9], where a more elegant optimization approach was applied. It is well known that the dependent variables $I_{gs}(t)$ and $I_{ds}(t)$ in a MESFET are nonlinear functions of $V_{gs}(t)$ and $V_{ds}(t)$, which can be represented by

$$I_{gs}(t) = FNL[V_{gs}(t),\ V_{ds}(t)] \tag{5.11}$$

$$I_{ds}(t) = FNL[V_{gs}(t),\ V_{ds}(t)] \tag{5.12}$$

Let us consider the device driven by a signal at a frequency ω. The device nonlinearity generates harmonic frequencies which can be expressed in terms of Fourier series as

$$V_{gs}(t) = V_{gs0} + \sum_{n=1}^{N} V_{gsn}\cos(n\omega t + \theta_n) \tag{5.13}$$

$$V_{ds}(t) = V_{ds0} + \sum_{n=1}^{N} V_{dsn}\cos(n\omega t + \varphi_n) \tag{5.14}$$

where, θ_n, φ_n are generic phases of the gate and drain waveforms.

Inserting (5.13) and (5.14) into (5.11) and (5.12), the resulting harmonics voltages and current coefficients at the external terminals are expressed by I_{1n}, I_{2n} and V_{1n}, V_{2n} respectively. The key to the optimization procedure is the definition of the objective function in terms of those vectors. For instance, for a power amplifier, a desired feature is the maximum added power at the output for a minimum input power with a low second harmonic content. The optimization function is then described by

$$P_{ad}(V_{gs}, V_{ds}) = \frac{1}{2}Re(V_{21}I_{21}^*) - \frac{1}{2}Re(V_{11}I_{11}^*) - \frac{1}{2}Re(V_{22}I_{22}^*) \tag{5.15}$$

where,

P_{ad} = represents the added power of the amplifier

V_{21} = represents fundamental frequency voltage at output terminal

I_{11} = represents fundamental frequency current at input terminal

The first term in the equation represents the fundamental frequency power generated at the output; the second term, the applied power at the input; and the third,the generated second harmonic power at the drain. During the optimization process,the variables are submitted to the following constraints:

- They must obey the nonlinear conditions (5.11) and (5.12);
- Their variations are limited by conditions allowing their physical realizability. During the optimization process, the variables from (5.15) are constrained to the physical realization of the networks. In other words, their variations are limited by passivity and activity of the coupling networks. For instance, the transistor must present a passive impedance

to the generator at the fundamental frequency, as expressed by (5.16) and operate as a generator at frequencies nω, expressed by (5.17).

$$Re[V_{11}I_{11}^*] > 0 \tag{5.16}$$

$$Re[V_{1n}.I_{1n}^*] + Re[V_{2n}.I_{2n}^*] < 0 \text{ for } n = 2,3, \ldots N \tag{5.17}$$

Once optimization is achieved, the terminal voltages and currents are known, and from them the input and output impedances given by

$$Z_g(\omega) = (V_{11}/I_{11})^* \qquad Z_g(2\omega) = V_{12}/I_{12} \tag{5.18}$$

$$Z_d(\omega) = (V_{21}/I_{21})^* \qquad Z_d(2\omega) = V_{22}/I_{22} \tag{5.19}$$

In the case of a frequency multiplier, the desired feature is a maximum power added at a frequency nω, for a minimum input power at a frequency ω. The objective function can be described by

$$P_{ad}(V_{gs}, V_{ds}) = \frac{1}{2}Re(V_{2n}I_{2n}^*) - \frac{1}{2}Re(V_{1n}I_{1n}^*) - \frac{1}{2}Re(V_{11}I_{11}^*) \tag{5.20}$$

The first term represents the average power at the drain at the output harmonic, and the second, the power generated by the harmonics at the input. The input power at the fundamental frequency is represented by the third term in the equation.

It is important to emphasize that the optimization is carried out without a predetermined coupling network using terminal voltages and currents as variables. Similar to the amplifier case, the harmonic impedances are determined from the resulting voltages and currents after the completion of the optimization process. The ratio between them will determine the device large-signal impedances that will be employed in the design of an optimum matching network.

5.6 Harmonic Load Pull Approach

5.6.1 Fundamental Frequency Load Pull

The first systematic application of load pull techniques was devoted to power amplifier design and has become a standard procedure [10] for other types of circuits. The experimental design steps are described in Figure 5.20 and the measurement set-up is shown in Figure 5.22.

The method is based on the assumption that, at the measured performance, which was manually adjusted, the tuner impedance $Z_L(\omega)$ is equal to

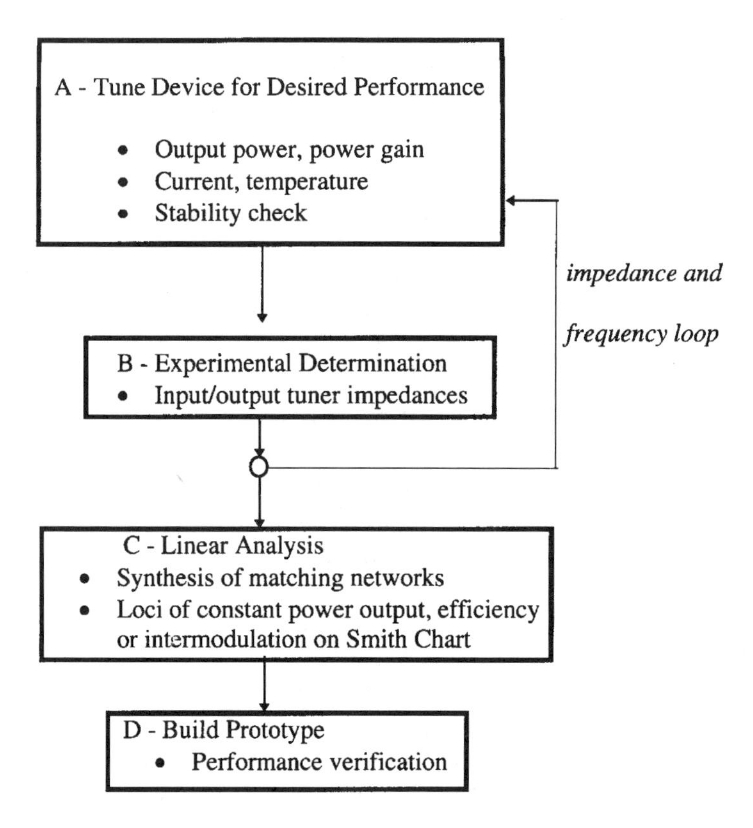

Figure 5.20 Load pull method.

$Z_D(\omega)^*$. Initially the device is tuned in a test fixture to specific operating conditions such as maximum output power. The input power, current, and temperature information are collected at a given frequency. Any tuner can be used, either coaxial slug tuners, waveguide screw tuners, or equivalent microstrip components. In the second step, the device is removed, the tuner impedance is measured on a network analyzer, and the resulting impedance is transferred to the device terminals, taking into account fixture and transmission-line losses and transition effects. The process continues by choosing a new impedance for a different power and the results are plotted on a Smith chart. Successively repeating this process, it is possible to determine loci of constant power output contours, constant efficiency, constant gain, intermodulation, or any other parameter of interest as a function of frequency.

The "modeled impedance" approach is found by the set of impedances for a desired output performance; for instance, the set corresponding to output power and the set corresponding to efficiency. A network is then synthesized to transform the 50 ohms load to the experimentally determined impedance.

In spite of the time involved in building up a reasonable amount of data to be useful for a wideband design, this method has been widely employed in the design of microwave circuits. The reason for its acceptance is the simple direct determination of potential devices under the conditions at which it is supposed to operate in the final circuit. Therefore, it provides information on power output, large-signal power gain, and large-signal input/output impedances. Typical power contours on a Smith chart are shown in Figure 5.21. The center of the ellipse shaped contour is the point corresponding to the maximum output power, and its impedance Z_{LM} can be read directly on the Smith chart. The smaller ellipse corresponds to the family of load impedances providing an output power 1 dB below the maximum. The greater ellipse corresponds to a power output 2 dB below the maximum. A similar contour can be obtained for a different test frequency and can be overlaid on top of the previous contour in the chart. That allows modeling the load impedance as a function of frequency.

In general, the measurement set-up requires the use of a sweeper generator, power meters, couplers, and other hardware. The spectrum analyzer is included in the measurement to detect any circuit instability during the tuning process. It can also be used to optimize the circuit for best intermodulation performance in the case of amplifiers. The device under test (DUT) is connected between the input and output tuners which will adjust the impedance to the required performance.

The accuracy of the load-pull method is limited by the accuracy with which the terminations and power levels can be determined. To minimize the errors the directional coupler, isolators, adapters, and test fixture used in the

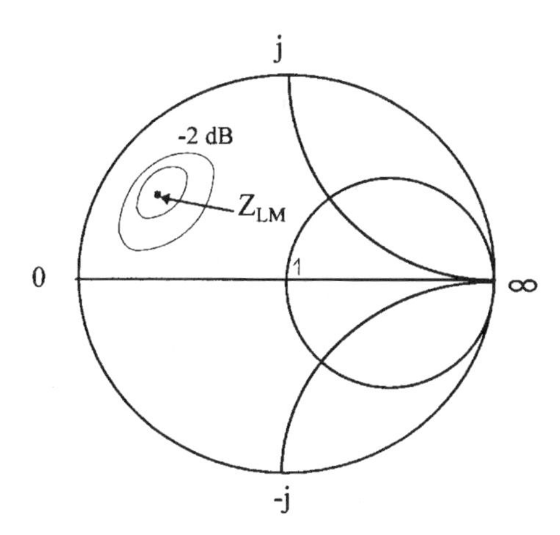

Figure 5.21 Loci of constant power as a function of impedance at frequency = f_0.

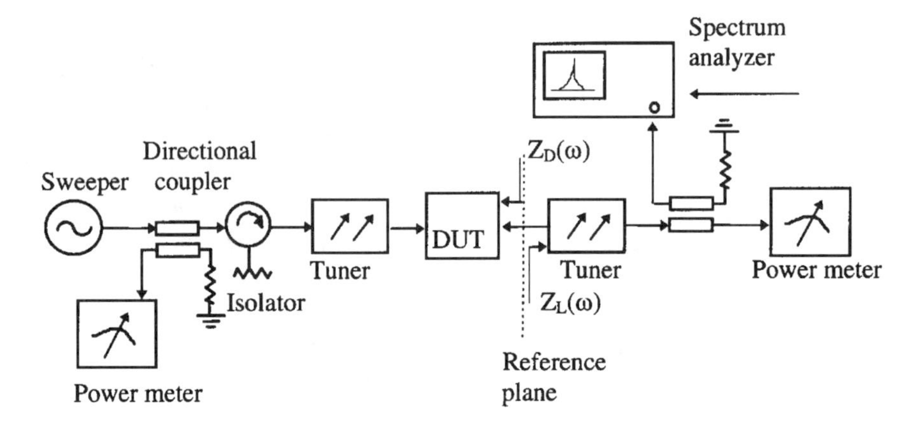

Figure 5.22 Basic amplifier load pull system.

test set-up must present low losses and a low VSWR. The tuners deserve special attention, since they must be capable of presenting a low VSWR in the untuned mode and also the high VSWR required for matching the device.

In power amplifiers, it may happen that a load required for high power operation is equal to 1 ohms, and that level of impedance can only be presented by the tuner if they have very low losses. The losses are also very important in the estimation of mismatch losses introduced in the measurement. For instance, the diagram of Figure 5.23 shows the ideal and real mismatch losses of a tuner.

The insertion loss, IL, of the ideal tuner is described by (5.21), a lossy (real) tuner by (5.22), and the relation between reflection coefficients and insertion loss by (5.23). L represents the tuner losses in dB.

$$IL(\text{dB}) = 10\log(1 - |\Gamma_1|^2) \tag{5.21}$$

$$IL(\text{dB}) = 10\log(1 - |\Gamma_1|^2\ 10^{L/5}) \tag{5.22}$$

$$\Gamma_2 = \Gamma_1(10^{L/10}) \tag{5.23}$$

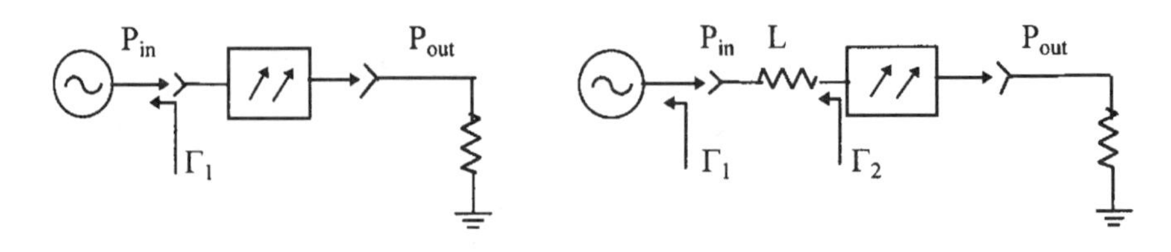

Figure 5.23 Mismatch losses (a) ideal tuner; (b) real tuner.

According to these equations, the 1 ohms impedance generated by the tuner, corresponding to $\Gamma_2 = 0.96$ will be modified at the device terminal due to the 1 dB losses to $\Gamma_2 = 0.96(10^{-1/10}) = 0.77$, and that corresponds to an impedance of 6.56 ohms.

5.6.2 Automatic Load Pull

Recently, automatic load pull systems [12] have appeared in the market where network analyzers are inserted between the device and the tuner to measure the ratio of incident and reflected power levels. A typical diagram from the RF part of such a system is depicted in Figure 5.24. The control unit, usually controlled by a PC computer is not shown. Therefore, impedance levels can be determined without the need to disassemble the test fixtures after each measurement, resulting in a much reduced time and higher accuracy compared to the manual method. The system is calibrated using a series of conventional small signal network analyzer calibration measurements.

Going one step further, such systems are equipped with software tools and computer-controlled, precision servomechanism systems capable of searching for the desired impedances. For instance, it can search for the reflection coefficients that will provide an approximately constant output power and will plot the results directly on the Smith chart. Obviously, other parameters can simultaneously be determined from such methodology.

5.6.3 Harmonic Load Pull

A harmonic load pull [12] system was introduced in the late seventies, where the conventional automatic load pull system was modified to include the possibility of independent fundamental frequency and second harmonic tuning. The set-up depicted in Figure 5.25 is designed specifically for frequency doublers. In this case, the coax to waveguide transition act as a highpass filter, where the fundamental frequency is reflected back to the device, and its phase

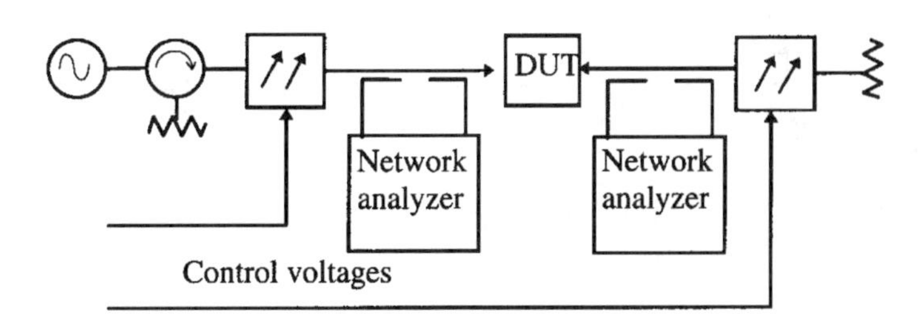

Figure 5.24 Basic diagram of an automatic load pull system.

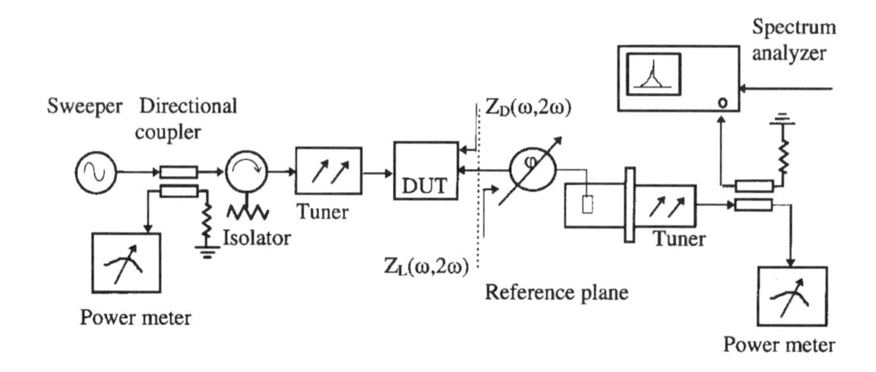

Figure 5.25 Basic multiplier load pull system.

is controlled by the phase-shifter. The second tuner on the waveguide performs tuning only on the output harmonic. The spectrum analyzer is connected at the output to check for oscillations, phase noise performance, or harmonic content. It can also be connected at the input during the measurement, to detect low frequency oscillations that are below the waveguide cutoff frequency.

Although this is not an automated system, it allows characterization of frequency doublers. Higher order multipliers usually generate a higher level of unwanted harmonics, requiring the use of additional bandpass filter at the output. This set-up is relatively inexpensive and presents reasonable accuracy. An automatic harmonic load pull shown in Figure 5.26 contains one network analyzer for the fundamental frequency signal and another for the second harmonic. The input tuning is similar to the arrangement of Figures 5.24 and 5.25. The distorted drain signal is applied to the harmonic tuner, the fundamental frequency is tuned by tuner 1, is reflected back to the device by the circulator action and the reactance of the highpass filter. The second harmonic frequency

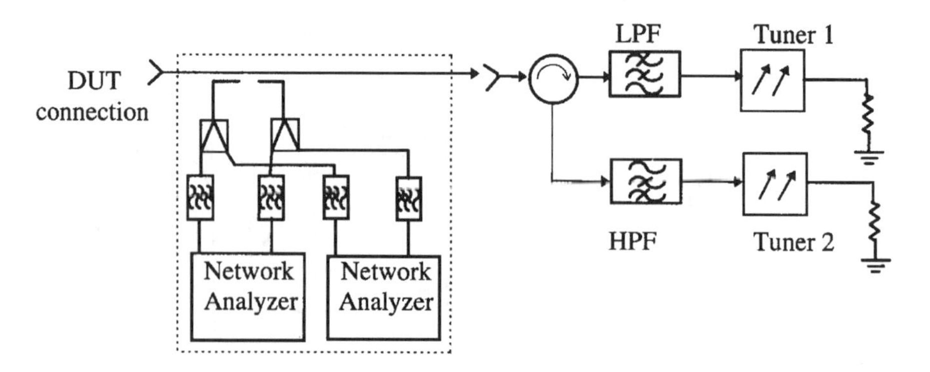

Figure 5.26 Harmonic load pull.

on the other hand, is reflected by the reactance of the lowpass filter and is tuned by tuner 2. By the circulator action, it is reflected back to the device. It is obvious that the circulator has to be broadband and present very low loss for generating a wide range of standing waves. Higher harmonics are essentially absorbed by the circulator, since they are out of band, and have no effect on the measurements.

This method is applicable to the characterization of power amplifiers and frequency doublers, but requires very careful calibration of the whole system. Otherwise, the desired results may be masked by inherent system errors. The losses introduced by the tuner may also be a limitation of the applicability of such procedure. To overcome these problems, active tuners [12] were introduced, where an amplifier is used to compensate the system losses. An example of this arrangement is shown in Figure 5.27. By properly adjusting the attenuator and the phase-shifter, it is possible to present at the device terminal almost any impedance located inside the Smith chart.

Observe that the amplifier is required to be of much higher power than the signals it will be amplifying, to avoid introduction of nonlinear effects in the measurement. The loop gain has to be carefully controlled to avoid instabilities generated by the power amplifier. In spite of such limitations, one can find in the market active harmonic automatic load pull systems with reasonable performance.

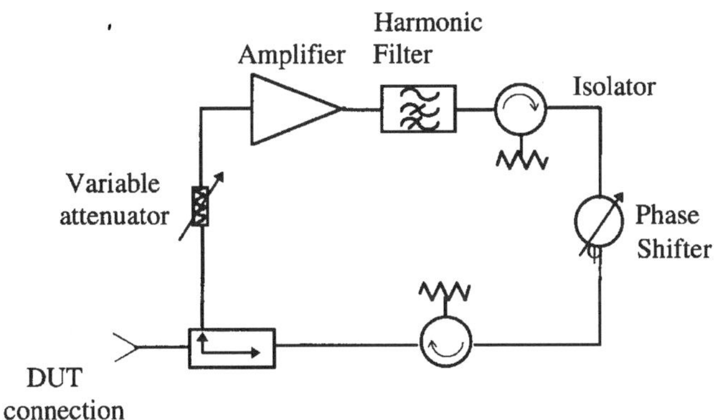

Figure 5.27 Active tuner load.

References

[1] Gelb, A., and W. E. Wander Velde, *Multiple-Input Describing Functions and Nonlinear System Design,* New York: McGraw-Hill Book Co, 1968.

[2] Rauscher, C., "High Frequency Doubler Operation of GaAs Field-Effect Transistors," *IEEE Transactions on Microwave Theory and Techniques,* Vol MTT-31, June 1983, pp. 462–473.

[3] *Pspice User's Guide,* Microsim Corp., October 1986.

[4] *TOUCHSTONE User Manual,* HP-EESOF, Westlake Village, CA.

[5] Rauscher, C., "Large-Signal Technique for designing single frequency and voltage controlled GaAs FET Oscillators," *IEEE Transactions on Microwave Theory and Techniques,* Vol MTT-29, April 1986, pp. 293–304.

[6] Abe, H., "A GaAs MESFET Oscillator Quasi-Linear Design Method," *IEEE Transactions on Microwave Theory and Techniques,* Vol MTT-34, No. 1, January 1986, pp. 19–25.

[7] *LIBRA User's Manual,* HP-EESOF, Westlake Village, CA.

[8] *MDS—Microwave Design System Users Manual,* HP-EESOF, Santa Rosa, CA, 1996.

[9] Guo, C., E. Ngoya, R. Quere, M. Camiade, and J. Obregon, "Optimal CAD of MESFETS Frequency Multipliers with and without Feedback," IEEE MTT-S 1988 International Microwave Symposium, New York, 1988, pp. 1115–1118.

[10] Belohoubek, E. F., A. Rosen, D. M. Stevenson, and A. Presser, "Hybrid Integrated 10-Watt CS Broad-Band Power Source at S Band," *IEEE Journal of Solid State Circuits,* Vol SC-4, December 1969, pp. 360–366.

[11] Cusack, J., M. S. Perlow, and B. S. Perlman, "Automatic Load Contour Mapping for Microwave Power Transistors," *IEEE Transactions on Microwave Theory and Techniques,* Vol MTT-22, No 12, December 1974, pp. 1146–1152.

[12] Stancliff, R., and D. D. Poulin, "Harmonic Load-Pull," IEEE MTT-S 1979 International Microwave Symposium Digest, 1979, pp. 185–187.

6

FET Harmonic Oscillators

Harmonic oscillators are components where one single device functions both as a fundamental frequency oscillator and as a harmonic generator. These components deserve special attention due to their capability of generating RF power above the frequencies usually obtained from a fundamental frequency oscillator. Their design requires simultaneous application of oscillator theory and frequency multiplier theory, in order to obtain a component with a reasonable performance in output power, bandwidth, and phase noise.

A simple means of introducing the trade-offs relative to a harmonic oscillator is to start out on thermodynamic balance of power [1] in a microwave amplifier, depicted in Figure 6.1. The amplifier can be visualized as a DC/RF power converter, whose total input/output powers are described by (6.1).

$$P_{in} + P_{DC} = P_{out} + P_{diss} \tag{6.1}$$

This equation can be modified to include the amplifier power gain, G, and the RF power added by the amplifier, P_{ad}, resulting in the following

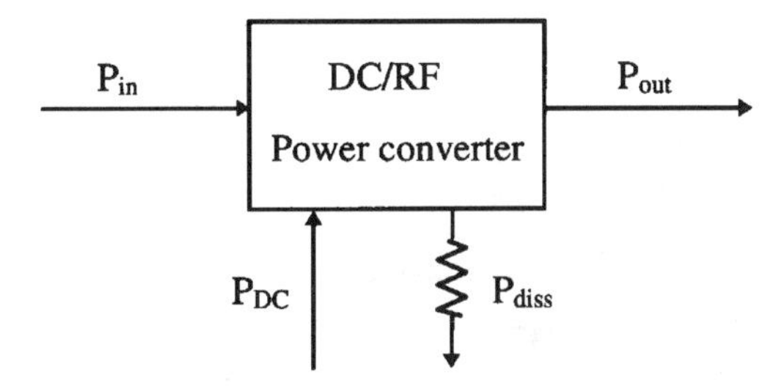

Figure 6.1 Thermodynamic representation of an amplifier.

$$P_{diss} = P_{DC} - (P_{out} - P_{in}) = P_{DC} - (G - 1)P_{in} \quad (6.2)$$

$$P_{ad} = P_{out} - P_{in} \quad (6.3)$$

The first equation shows that if the power gain remains constant, then a continuous increase in input power will increase the second term of the equation until P_{diss} becomes negative. Since this is physically impossible in thermodynamics (assuming the dc power remains constant), one concludes that under large-signal drive, the amplifier gain G has to decrease.

In the case of an oscillator, the situation is similar; part of the drain energy is fed back to the gate through a coupling network in order to sustain oscillations. The previous diagram is therefore modified to take into account the feedback depicted in Figure 6.2.

The power balance for the oscillator is also similar to the amplifier case and can be rearranged in the form defined by (6.4), where P_{osc} represents the power delivered to the load. Conventional oscillator theory states that the maximum available power from an oscillator is equal to the added power from an amplifier at the point of maximum efficiency (i.e. when the difference between P_{out} and P_{in} is maximum).

$$P_{osc} = P_{out} - P_{in} = P_{DC} - P_{diss} \quad (6.4)$$

$$P_{osc} = P_{ad} = \max\{P_{out} - P_{in}\} \quad (6.5)$$

The harmonic oscillator configuration is actually the same as Figure 6.2 with the addition of a harmonic filter at the output. Now the output power comprises fundamental frequency power P_{of} plus harmonic power P_{oh}. In (6.5) it is then transformed to

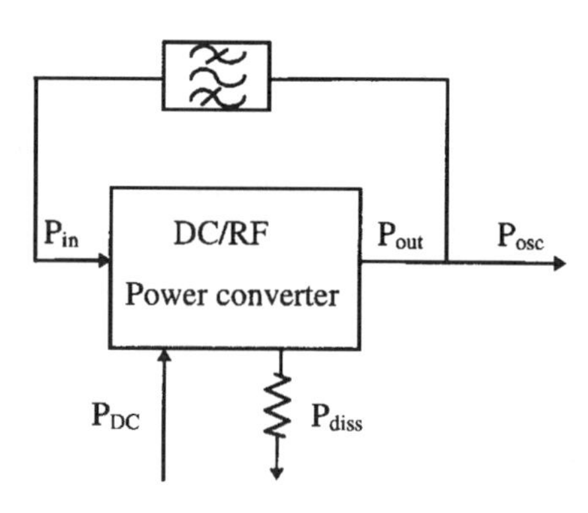

Figure 6.2 Thermodynamic representation of an oscillator.

$$P_{hosc} = (P_{of} - P_{in}) + P_{oh} = (P_{oh} - P_{in}) + P_{of} \quad (6.6)$$

This equation shows that an optimum harmonic oscillator should generate power at the fundamental frequency just enough to sustain oscillations, most of the output power should be delivered to the external load and the difference will be dissipated in the other harmonics. Now the condition for maximum harmonic power will take place when the difference $(P_{oh} - P_{in})$ is maximum.

6.1 Design Approach

The design of harmonic oscillators starts with the design of a frequency multiplier, employing any of the methods described in the previous chapter. Then, the conventional design requirements for oscillator operation are modified to match the frequency multiplier requirements.

The objectives of a frequency multiplier are a maximum output power at the desired harmonic, a matched input for maximum efficiency, and a reasonable frequency bandwidth. For harmonic oscillator operation, the multiplier's saturation characteristics are also required, which can be obtained from experimental measurements or through nonlinear simulation. The parameters sought are the ones indicated by

$$P_{oh}(P_{in}, n\omega_0) - P_{in}(\omega_0) = f[P_{in}(\omega_0)] \quad (6.7)$$

This is a common procedure in oscillator design [2], and is equally applicable to harmonic oscillators. The simulation results for a 5 to 10 GHz frequency doubler are shown in Figure 6.3, employing the reference device. Similar relations are obtained for higher order multipliers.

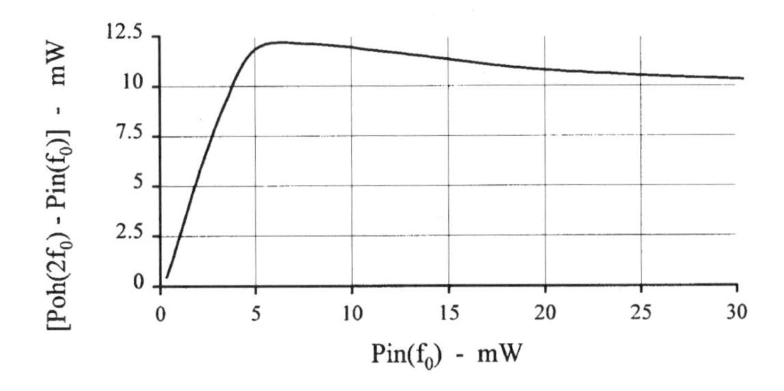

Figure 6.3 Saturation characteristics of a frequency multiplier.

One can observe a nonlinear relation between output power and input drive at low levels, and a saturation of the second harmonic output power after a certain input drive level. The added power is maximized at the peak of the curve, obtained for a matched input that maximize the gate voltage swing over the control capacitor, a shorted load for the fundamental frequency maximizing the drain current, and a matched second harmonic drain impedance.

In the case of harmonic oscillators, (6.6) has to be taken into consideration, where the frequency multiplier is modified to include generation of power at the fundamental frequency enough to sustain oscillations, (i.e. $P_{out}(P_{in}, \omega_0) = P_{in}(\omega_0)$). Therefore, the drain circuit has to be modified so that the drain power at the fundamental frequency is equal to the power absorbed by the gate. An accurate procedure for the design of the drain impedance at the fundamental frequency is to monitor both the gate and drain power at each iteration, using (6.8).

$$P_{av} = Re\{V(\omega).I(\omega)^*\}/2 \tag{6.8}$$

The nonlinear analysis will provide a set of voltages and currents at the fundamental frequency and harmonic frequencies, which can be used to generate the feedback network. The currents in the multiplier are obtained using the receptor convention, where the currents are defined entering at gate and drain ports. They must be modified to the generator condition, where they leave the gate and drain ports, as depicted in Figure 6.4.

The next step is to insert the device and associated I,V into any of the standard oscillator coupling topologies available, either "T" or "PI," illustrated in Figures 6.5 and 6.6, respectively. The network elements must satisfy the I,V oscillation conditions at the fundamental frequency [2] and the harmonic conditions established by the frequency multiplier prototype. Three options are possible for each topology, with the load connected either to the drain, to the source, or the gate. The connection of the load to the drain or source provides more power compared to the gate connection. The reason for higher power is due to the higher voltage and current available on the output plane.

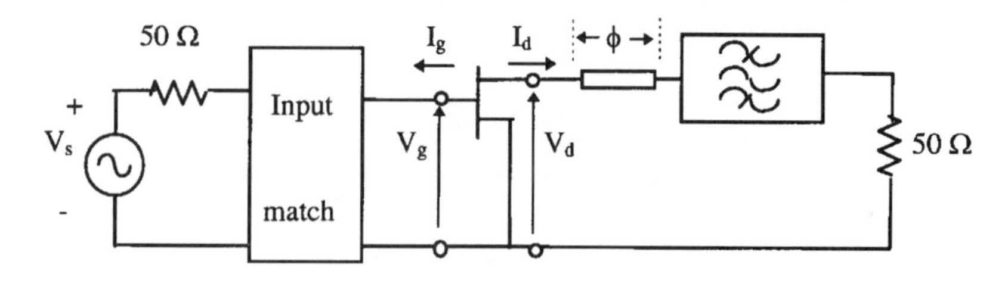

Figure 6.4 Simplified doubler circuit.

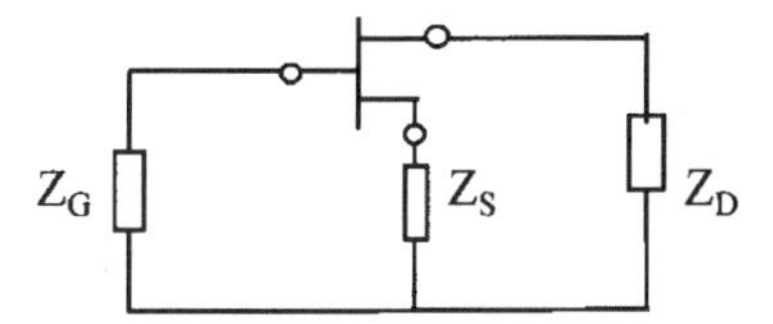

Figure 6.5 "T" topology for the harmonic oscillator.

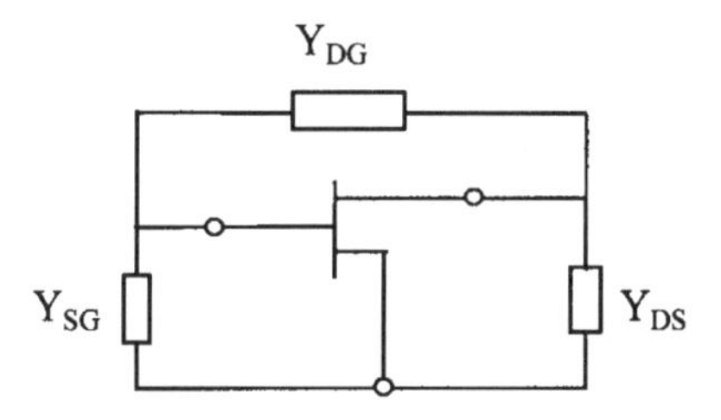

Figure 6.6 "PI" topology for the harmonic oscillator.

If the dc power requires grounding the source, then there is only one choice. In the case of low power oscillators, there is no preferred connection other than layout convenience. The examples in the figures are for a load connected to the drain.

The terminations described in Table 6.1 are for the fundamental frequency and output harmonic only using the "T" topology. The terminations for other harmonics should be selected to enhance the desired drain waveform.

The boundary I,V conditions have to be valid simultaneously at the fundamental frequency and at the harmonics. The set of harmonic impedances from Table 6.1 is preferred, where the source impedance is shorted to ground. The terminating impedances are defined by the I,V relationships. The set of terminations for a "PI" topology are described in Table 6.2.

The equations required to calculate the network elements as a function of I,V at the fundamental frequency are in Tables 6.3 and 6.4. Note that the

Table 6.1
Network Elements for the "T" Topology

Fundamental Frequency Impedances	**Output Harmonic Frequency Impedances**
$Z_D(\omega_o) = R_{D1} + jX_{D1}$	$Z_D(n\omega_o) = R_{Dn} + jX_{Dn}$
$Z_S(\omega_o) = jX_{S1}$	$Z_S(n\omega_o) = 0$
$Z_G(\omega_o) = jX_{G1}$	$Z_G(n\omega_o) = R_{Gn} + jX_{Gn}$

Table 6.2
Network Elements for the "PI" Topology

Fundamental Frequency Admittance	Output Harmonic Frequency Admittance
$Y_{DS}(\omega_0) = G_{DS1} + jB_{DS1}$	$Y_{DS}(n\omega_0) = G_{DSn} + jB_{DSn}$
$Y_{GS}(\omega_0) = jB_{GS1}$	$Y_{GS}(n\omega_0) = G_{GSn} + jB_{GSn}$
$Y_{DG}(\omega_0) = jB_{DG1}$	$Y_{DG}(n\omega_0) = 0$

Table 6.3
"T" Configuration: Network Elements as a Function of Terminal I,V at the Fundamental Frequency [2]

Case	Z_G	Z_S	Z_D
$Re\{Z_G\} = Re\{Z_S\} = 0$	$\frac{jRe\{V_G(I_G^* + I_D^*)\}}{Im\{I_G^* I_D\}}$	$\frac{jRe\{V_G I_G^*\}}{Im\{I_G I_D^*\}}$	$\frac{V_D}{I_D} - Z_s\left[1 + \frac{I_G}{I_D}\right]$
$Re\{Z_G\} = Re\{Z_D\} = 0$	$\frac{jRe\{I_D(V_D^* - V_G^*)\}}{Im\{I_G I_D^*\}}$	$\frac{V_G - Z_G I_G}{I_G + I_D}$	$\frac{jRe\{I_G(V_G^* - V_D^*)\}}{Im\{I_G I_D^*\}}$
$Re\{Z_D\} = Re\{Z_S\} = 0$	$\frac{V_G}{I_G} - Z_s\left[1 + \frac{I_D}{I_G}\right]$	$\frac{jRe\{V_D I_D^*\}}{Im\{I_G^* I_D\}}$	$\frac{jRe\{V_D(I_D^* + I_G^*)\}}{Im\{I_G I_D^*\}}$

Table 6.4
"PI" Configuration: Network Elements as a Function of Terminal I,V at the Fundamental Frequency [2]

Case	Y_{GS}	Y_{GD}	Y_{DS}
$Re\{Y_{GS}\} = Re\{Y_{GD}\} = 0$	$\frac{jRe\{I_G(V_D^* - V_G^*)\}}{Im\{V_D V_G^*\}}$	$\frac{jRe\{V_G^* I_G\}}{Im\{V_G^* V_D\}}$	$\frac{I_D}{V_D} + Y_{GD}\left[\frac{V_G}{V_D - 1}\right]$
$Re\{Y_{GS}\} = Re\{Y_{DS}\} = 0$	$\frac{jRe\{V_D(I_D^* + I_G^*)\}}{Im\{V_D V_G^*\}}$	$\frac{Y_{GS} V_G - I_G}{V_D - V_G}$	$\frac{jRe\{V_G(I_G^* + I_D^*)\}}{Im\{V_D^* V_G\}}$
$Re\{Y_{GS}\} = Re\{Y_{DS}\} = 0$	$\frac{I_G}{V_G} - Y_{GD}\left[\frac{V_D}{V_G} - 1\right]$	$\frac{jRe\{V_D^* I_D\}}{Im\{V_D^* V_G\}}$	$\frac{jRe\{I_D(V_G^* - V_D^*)\}}{Im\{V_D^* V_G\}}$

resulting drain load at the fundamental frequency is low, and represents the power required by the gate circuit to sustain oscillations.

Although both topologies are possible from the point of view of oscillators, the "T" configuration is more adequate to harmonic oscillators, due to the difficulty of building a circuit where the feedback element Y_{DG}, is an open-circuit at the output harmonic.

6.2 Small-signal Check

The described nonlinear design approach is applied at the frequency of oscillation, ω_0, using nonlinear simulation, and provides a direct synthesis of the network required to build a harmonic oscillator. However, the method uses compressed device parameters, which are different from small-signal conditions, where the noise excites the negative resistance at the moment the circuit is switched on. Designing a circuit based only on large-signal conditions alone does not guarantee the start up of oscillations. Therefore, a check of the small-signal oscillation condition is a complementary step to the large-signal approach.

The analysis from the impedance point of view [3], starts by breaking the circuit into two parts, one containing the active device and presenting an impedance $Z_D(\omega_0, V)$ which is dependent on frequency and signal amplitude, V, and the other containing the resonator whose impedance $Z_L(\omega_0)$ is only dependent on frequency. The circuit on Figure 6.7 depicts a break at the gate connection.

The start-up oscillation conditions are given by (6.9) and (6.10), where the net resistance is negative. Thus, when the circuit is turned on, the transients or the white noise voltages incident in the resistance reflects back with higher amplitude for the frequencies within the band of the resonator.

$$Re(Z_L(\omega_0)) + Re(Z_D(\omega_0, V)) < 0 \quad (6.9)$$

$$Im(Z_L(\omega_0)) + Im(Z_D(\omega_0, V)) = 0 \quad (6.10)$$

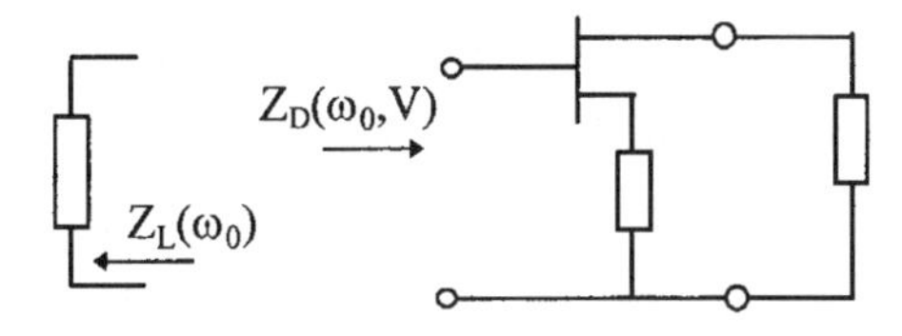

Figure 6.7 Breaking the circuit into two components for small-signal check.

The stabilized oscillation conditions are met when a particular steady-state oscillation frequency and voltage amplitude is such that condition (6.9) becomes equal to zero, or

$$Re(Z_L(\omega_0)) + Re(Z_D(\omega_0, V)) < 0 \tag{6.11}$$

Therefore, at start-up the net resistance is negative, and as the signal amplitude increases, the device impedance modifies until it matches the resonator impedance at steady-state. This condition is applicable when the device impedance behaves as a series RLC circuit, (i.e., when the reactance slope is positive, or $dX/d\omega > 0$). If the device impedance behaves like a parallel RLC circuit, (i.e., when the susceptance slope is positive), then the admittance approach should be used.

A third approach to analyze the oscillation conditions, consists of describing the impedance or admittance in terms of reflection coefficient from the resonator, Γ_L, and the active device, Γ_D, described by

$$\Gamma_L = (Z_L - Z_0)/(Z_L + Z_0) \tag{6.12}$$

$$\Gamma_D = (Z_D - Z_0)/(Z_D + Z_0) \tag{6.13}$$

The oscillation condition is then expressed by an equation similar to (6.9), (6.10),

$$|\Gamma_D(\omega_0)|.|\Gamma_L(\omega_0)| \geq 1 \tag{6.14}$$

$$\text{and } \angle\Gamma_D(\omega_0) = \angle\Gamma_L(\omega_0) \tag{6.15}$$

This approach may predict a different oscillation frequency compared with the previous methods [3], which is demonstrated by (6.16) to (6.19).

$$Z_D = -R_D + jX_D \tag{6.16}$$

$$Z_L = R_L + jX_L \tag{6.17}$$

Let us assume the oscillation conditions are given by: $X_L = -X_D$, and $R_L = |R_D/3|$. Calculating the respective reflection coefficients, one obtains the phases given by (6.18) and (6.19), which obviously does not satisfy 6.15, since $\angle\Gamma_L \neq \angle\Gamma_D$.

$$\angle\Gamma_D = \frac{2X_D(R_D - Z_0)}{(R_D - Z_0)^2 + X_D^2} \tag{6.18}$$

$$\angle \Gamma_L = \frac{-2X_L(R_L - Z_0)}{(R_D/3 - Z_0)^2 + X_L^2} \tag{6.19}$$

Therefore, the start-up conditions predicted by the reflection approach are different from the ones predicted by either impedance or admittance. However, when steady-state oscillations are established, all three approaches are coincident. In spite of the drawback presented by the reflection coefficient for the start-up of oscillations, it is a very convenient tool in oscillator design, allowing a visual verification of the resonator and device impedance interaction on the Smith chart over a wide band. The region outside the Smith chart represents negative resistance, or in terms of reflection coefficient, $|\Gamma| > 1.0$. However, a simple transformation taking the conjugate inverse of the device reflection coefficient, causes the negative resistance to fall inside the positive Smith chart. The reactance values read from the chart are still the same, but the resistive ones need to be multiplied by -1 to obtain the correct value. Therefore, the passive impedance of the tank circuit can be overlaid on top of the active impedance, easing the study of oscillation condition.

An example of such analysis is shown in Figure 6.8 where the trace of the tank circuit impedance encircles the trace of the active circuit impedance. When laying out these plots it is important to note the electrical properties of each plot. The passive circuit plot is derived from a RLC parallel tank circuit, which presents a fast varying function of frequency, $\Gamma_L(\omega)$ with a self resonance at ω_0. This is a typical characteristic of any high Q-resonator. The device circuit plot is typical for a common source oscillator, whose small-signal reflection coefficient $\Gamma_D(\omega, V)$, is a slow-varying function of frequency.

The stability condition for an oscillator is derived from perturbation theory [4] and is given by

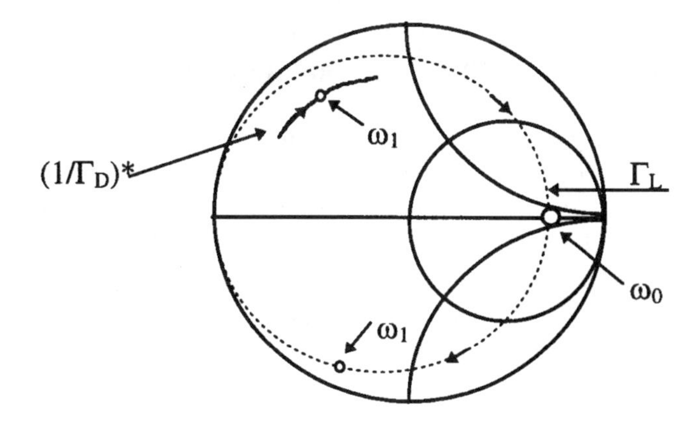

Figure 6.8 Analysis of oscillation conditions.

$$\left\{\left[\frac{\delta R_D}{\delta V}\right]_{V=V_0}\left[\frac{\delta X_L}{\delta \omega}+\frac{\delta X_D}{\delta \omega}\right]_{\omega=\omega_0}-\left[\frac{\delta RL}{\delta \omega}\right]_{\omega=\omega_0}\left[\frac{\delta X_D}{\delta V}\right]_{V=V_0}\right\}>0 \quad (6.20)$$

where V_0 and ω_0 represent the steady-state amplitude and frequency, respectively. This equation guarantees a stable operating point after the oscillations have grown from small to large signal conditions. For GaAs FETs, the following approximations can be made to simplify 6.20. As the signal grows and reaches the device compression point, the gain decreases and so does the negative resistance. Therefore,

$$(\delta R_D/\delta V)|_{V=V_0} < 0 \quad (6.21)$$

The device reactive elements are basically amplitude independent, so that

$$(\delta X_D/\delta V)|_{V=V_0} \cong 0 \quad (6.22)$$

Under these conditions, (6.20) reduces to (6.23):

$$\left[\frac{\delta X_L}{\delta \omega}+\frac{\delta X_D}{\delta \omega}\right]_{\omega=\omega_0} > 0 \quad (6.23)$$

The above equation states that the load and source reactive components must exhibit a positive slope about the operating point to ensure stable operation. Thus, if the device presents a series RLC characteristic, the tank circuit should also present the same characteristic.

6.2 Application to a 10 GHz Harmonic Oscillator

The objective is to design a second harmonic oscillator at the fundamental frequency of 5 GHz. The first step in the design is to define the device bias. For best second harmonic generation, it should be biased near the pinch-off gate voltage, V_P. However, biasing at pinch-off results in low transconductance which results in low loop gain, not adequate for starting oscillations [5]. Thus, in order to guarantee the build up of oscillations, the device should be biased at a voltage greater than V_P, in class A mode, resulting in a high value for the transconductance, as shown by the bias point A in Figure 6.9.

However, this gate bias results in lower second harmonic generation degrading the doubler efficiency. Employing a self-bias approach (point B)

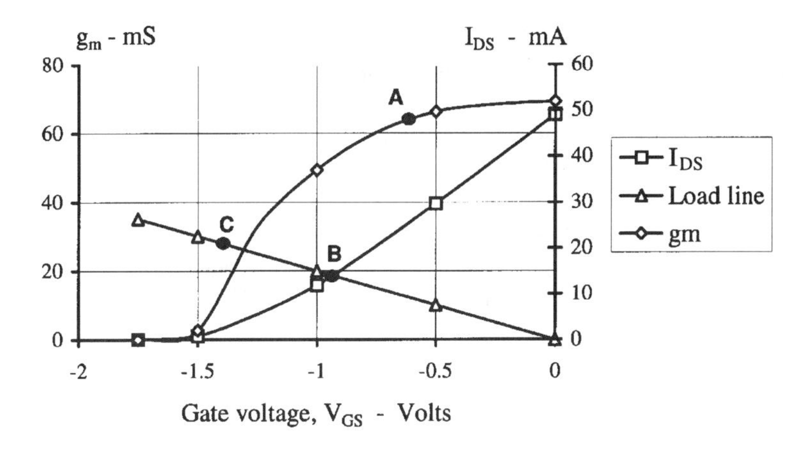

Figure 6.9 Self-bias in harmonic oscillators.

results in improved operation, since the increase of the RF signal generated at the start-up of oscillations moves the gate dc level to more negative values (point C), and stabilizes the bias at a point near the pinch-off voltage. The input dc load line that will provide this operation is also presented in the figure and was obtained assuming a half wave sinusoid for the drain current. Thus, one point on the line is $V_{GS} = 0$; $I_{DS} = 0$, and the other point is equal to $V_{GS} = V_P$; $I_{DS}/\pi \approx 60/\pi \approx 20$ mA. The drain bias is determined as the mid-point between the maximum drain voltage, BV_{DS}, and the saturation voltage, V_k.

The second step is to simulate a frequency doubler in order to determine the I,V set that will produce to best multiplier efficiency. The equivalent multiplier circuit is represented in Figure 6.10.

This circuit provides 6 dB gain with a return loss of 8 dB at 5.0 GHz, using the reference device in the design. The parameters for this circuit are represented in Table 6.5. The multiplier saturation characteristics were simu-

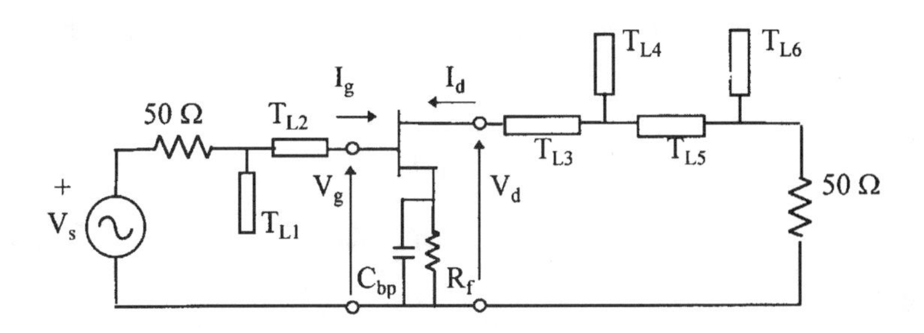

Figure 6.10 Simplified doubler circuit.

Table 6.5
Frequency Doubler Network Elements Electric Angles Normalized at f_0 = 5.0 GHz

Element No	Element Type	Parameters
T_{L1}	Open stub	$Z_0 = 25\Omega/\phi = 35.66°$
T_{L2}	Series line	$Z_0 = 85\Omega/\phi = 69.34°$
T_{L3}	Series line	$Z_0 = 62.5\Omega/\phi = 19.8°$
T_{L4}	Open stub	$Z_0 = 50\Omega/\phi = 75.3°$
T_{L5}	Series line	$Z_0 = 50\Omega/\phi = 110.9°$
T_{L6}	Open stub	$Z_0 = 50\Omega/\phi = 95.1°$
R_f	Resistor	R = 60 ohms
	C_{bp}	C = 20 pF

lated for an input drive from 0 to +15 dBm. The results are given in Figure 6.3 which shows a power-added of 12 mW for an input power of 5 mW.

The power balance for this circuit shows a gate power of 3.1 mW (5 dBm), a fundamental frequency drain power of 0.64 mW (–2 dBm) and a second harmonic power of 17.3 mW (12.4 dBm). The voltages, currents and respective harmonic powers are presented in Table 6.6, where the average power was calculated by means of equation (3.33):

From these results, the following notes should be made:

1. The gate power is equal to 5 dBm, 1 dB less than the available power from the generator. That difference is due to the imperfect input match, which introduces around 0.7 dB mismatch losses.
2. The drain power at the fundamental frequency is 0.6 mW, accounting for losses in the drain circuit. However, it is lower than the gate power and oscillations cannot be sustained if that power is fed back to the drain. Perform new iteration with a drain resistor that simulates the power required by the gate. Thus, $\Delta P = P_D - P_G = 2.5$ mW. The resistor is calculated assuming the drain current remains the same,

Table 6.6
I,V Set for the Frequency Doubler

Freq(GHz)	I_g (mA)	V_g (Volts)	P_G (mW)	I_d (mA)	V_d (Volts)	P_D (mW)
5.0	25.0/118.6°	2.36/34.5°	3.1	49.0/21.7°	0.53/–70.4°	0.64
10.0	2.3/111.6°	0.23/–156.4°	—	23.0/8.8°	1.70/–137.5°	17.3

then, $R = 2\Delta P / I_d^2 = 2.5$ ohms. The new set of I,V and the resulting power at the harmonics are given in Table 6.7.

This I,V set was applied to the design equations for a "TEE" type of oscillator at the fundamental frequency represented in Figure 6.5. The resulting impedances are presented in Table 6.8, from which the terminating impedances must be synthesized.

The device requires a slightly capacitive impedance in the source, and both the drain and gate should be terminated by inductive impedances. After connecting those impedances at the three terminals, it is possible to check the small-signal oscillation condition expressed by (6.9) and (6.10), looking into the gate and into the resonator circuit.

$$Z_G(5\text{GHz}) = -18.0 - j93.5\ \Omega$$
$$Z_L(5\text{GHz}) = 0.0 + j93.5\ \Omega$$

It is therefore guaranteed that the circuit will oscillate at the desired frequency of 5.0 GHz. The circuit of Figure 6.11 was designed to fit to the values of Table 6.8 at the fundamental frequency, and a best effort was dedicated to meet the second harmonic impedance.

Table 6.7
Modified I,V to Set $P_D > P_G$.

Freq (GHz)	I_g (mA)	V_g (Volts)	P_G (mW)	I_d (mA)	V_d (Volts)	P_D (mW)
5.0	25.0/119.0°	2.36/35.0°	3.1	48.0/21.7°	0.53/−83.8°	3.40
10.0	2.2/113.3°	0.22/−154.7°	—	23.0/10.6°	1.70/−135.9°	16.3
15.0	0.5/110.9°	0.02/14.0°	—	2.60/−96.3°	0.25/155.6°	0.10
20.0	0.2/−157.5°	0.05/3.2°	—	1.54/−49.2°	0.30/48.6°	—

Table 6.8
Terminating Impedances for the Harmonic Oscillator

Fundamental Frequency Impedances	Second Harmonic Frequency Impedances
$Z_D(\omega_0) = 0.18 + j15.48$	$Z_D(2\omega_0) = 61.3 + j40.8$
$Z_S(\omega_0) = 0.0 - j5.78$	$Z_S(2\omega_0) = 0.0$
$Z_G(\omega_0) = 0.0 + j97.7$	$Z_G(2\omega_0) = 3.5 + j100$

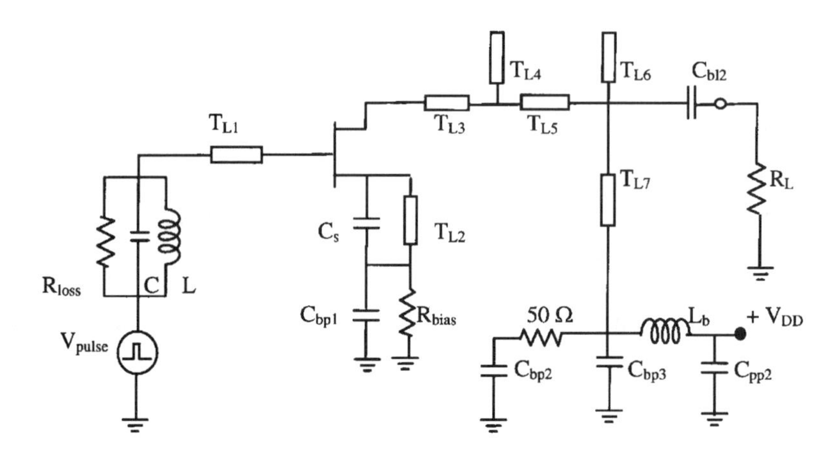

Figure 6.11 Schematic diagram of harmonic oscillator.

At the source, the capacitance required for oscillation, C_s, is short-circuited by a quarter-wavelength long transmission line at the second harmonic. The capacitor, C_{bp1}, shorts the bias resistor R_{bias} at the fundamental frequency and second harmonic.

The gate contains a tank circuit, whose impedance is adjusted by transmission line T_{L1} to present the required reactance at the fundamental frequency and to approximately meet the gate impedance at the second harmonic.

The same bandstop filter employed in the frequency multiplier, T_{L4} to T_{L6}, is used to block the fundamental and present a low loss at the second harmonic. The line T_{L3} provides the drain impedance from Table 6.8. The drain bias circuit is similar to the one used in Chapter 5.

The circuit was simulated using time domain techniques (Pspice) to determine power output, frequency and harmonic content. It has already been mentioned that the impetus to start-up oscillations comes from noise sources contained in the circuit. Since the frequency domain behavior of noise is spectrally flat for the frequencies under consideration, the time domain equivalent impetus can be obtained by the inverse Fourier transform and the result is a Dirac impulse waveform. A practical way to implement this perturbation signal in the circuit is to insert a narrow pulse generator, at a convenient point in the circuit. The circuit in Figure 6.11 shows its connection in series with the tank circuit.

A sample of the voltages and currents in the circuit are presented in Figures 6.12 to 6.15. It should be noted that none of the device terminals are grounded, and the voltages presented are with reference to ground. The first set in Figure 6.12 illustrates three complete cycles of the drain current and drain voltage. The transient effects occur before t = 19 ns, and are not shown

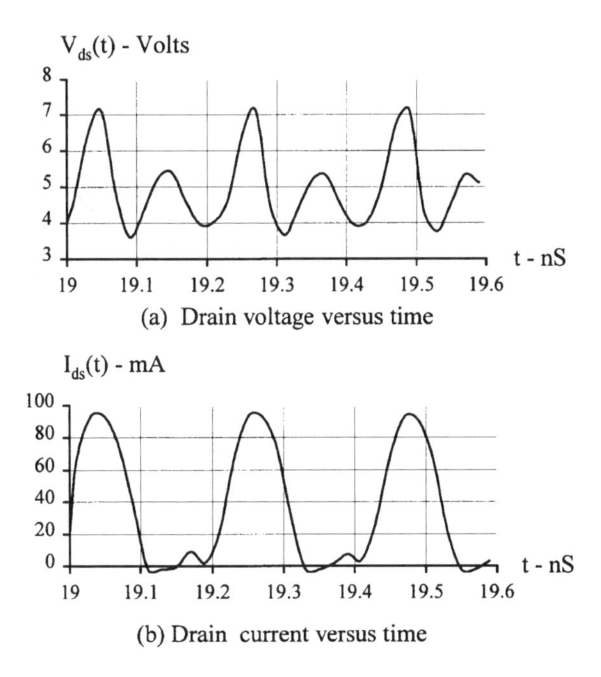

Figure 6.12 Representation of waveforms at the drain; (a) voltage and (b) current.

for the high number of cycles required to reach steady-state. The drain current appears as a half-wave rectified current. The drain voltage shows two components, the fundamental frequency component and the second harmonic component. The negative current observed on the negative cycles is the result of the numerical analysis and has no physical meaning.

The second set containing two complete cycles of the gate and drain current is shown in Figure 6.13. The gate current displays harmonic distortion due to the variation of input impedance within the signal period due to nonlinear drain current circulating in the drain circuit. The effect of employing a resonator with a high "Q" is observed on the gate voltage waveform, which short-circuited all gate current harmonics, displaying a very low harmonic content. During simulation, the losses, represented by resistor R_{loss}, had to be added to the resonator circuit and to the drain terminal in order to avoid convergence problems during the time domain simulation.

The representation of the load voltage in Figure 6.14, where the multiplication effect is observed by comparing the period at the load with the period at the gate in Figure 6.13. The period of the load voltage is half the period of the gate voltage (i.e., a frequency twice the oscillating frequency).

The voltage at the load shows very low harmonic content. A better measure of its "purity" is obtained by checking its spectrum, which is shown

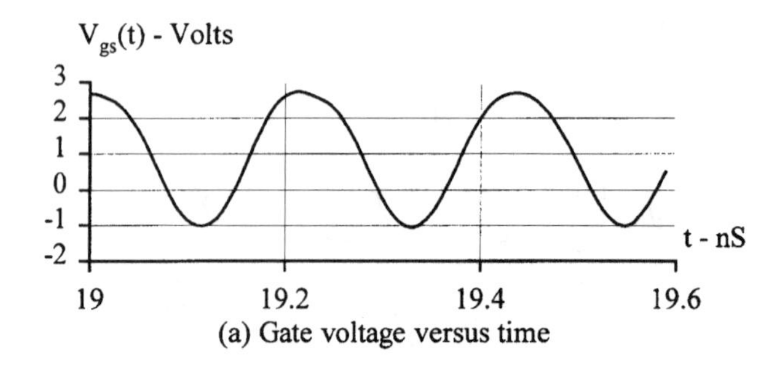

(a) Gate voltage versus time

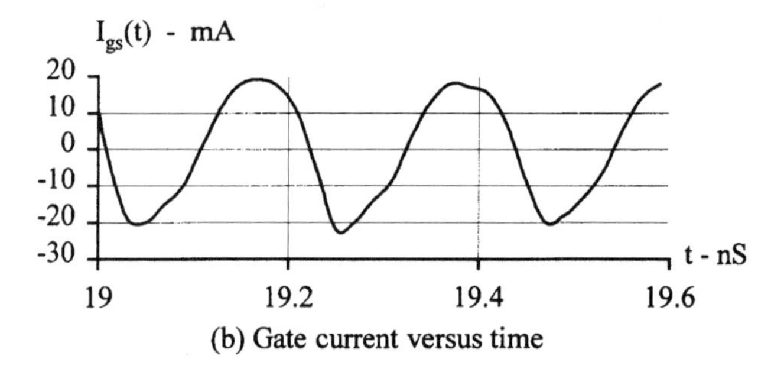

(b) Gate current versus time

Figure 6.13 Representation of gate waveforms; (a) gate voltage; (b) gate current.

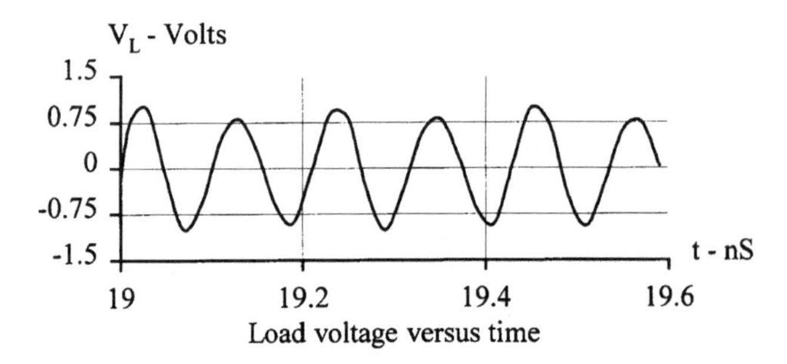

Load voltage versus time

Figure 6.14 Load voltage on the harmonic oscillator.

in Figure 6.15. The second harmonic at 10 GHz shows an output power in the order of +10 dBm. Compared to the output signal, the fundamental frequency component at 5 GHz, is 30 dB below the second harmonic, and the third is in the order of 25 dB below. The other components are of much less amplitude.

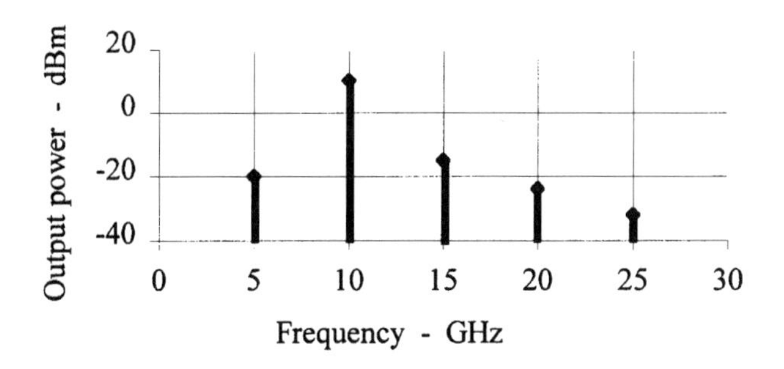

Figure 6.15 Spectrum at the output load.

References

[1] Soares, R., J. P. Castelletto, P. Legaud, and M. Armand, "Performance of An InP MISFET X-band Oscillator and K-band Oscillator-Doubler," European Microwave Conference, pp. 263–267.

[2] Rauscher, C., "Large-Signal Technique for Designing Single-Frequency and Voltage-Controlled GaAs FET Oscillators," *IEEE Transactions on Microwave Theory and Techniques,* Vol. 29, No. 4, April 1981, pp. 293–304.

[3] Savaria, S., and P. Champagna, "Linear Simulations for use in Oscillator Design," *Microwave Journal,* May 1985, pp. 98–105.

[4] Obregon, J. J., "Contribution a La Conception et a La Réalisation de Dispositifs Actifs Micro-Ondes a L'État Solide," These d'État Universite de Limoges, France, 1980.

[5] Tupynamba, R., E. Camargo, and F. S. Correra, "A HEMT Harmonic Oscillator Stabilized by an X-Band Dielectric Resonator," *MTT-S 1991 Intenational Microwave Symposium Digest,* Boston, MA, June 1991, pp. 277–280.

7

Typical Frequency Multiplier Topologies

The important performance parameters of a frequency multiplier are its multiplication gain, input return loss, and output power under stable conditions of operation. These conditions may be optimized by fine tuning the input and output circuits to achieve the desired terminating impedances at spot fundamental and harmonic frequencies. In general, this approach results in a narrowband multiplier, exhibiting less than 10% fractional bandwidth. Pursuing this approach at larger bandwidths becomes prohibitive, especially when the high end of the input frequency starts overlapping the low end of the output band. To design large bandwidth multipliers, the input and output device Q-factors have to be minimized by resistive loading, which also contributes to the circuit stability over the band. Then the appropriate topology is selected to provide the desired performance. For instance, inserting the FET in balanced topologies which inherently rejects the fundamental and odd harmonics at the output and employing MMIC technologies, octave bandwidth multipliers can be obtained.

In this chapter, several typical multiplier topologies are described and simulated performance are detailed for a selected set of circuits. The simulation includes RF parameters and waveforms for the voltages and currents. They are intended for either high efficiency, large bandwidth or high output power. The Statz-Pucel model of Chapter 2 was used to describe the chip device employed in all simulations. The application of MMIC technology in the design of multipliers is also covered with practical examples. The chapter is complemented with a few topologies of harmonic oscillators.

7.1 Frequency Doublers

Frequency doublers are the most common application of frequency multipliers due to their high efficiency, circuit simplicity, and minimum generation of unwanted harmonics. Most of the MIC applications found in the literature are for single-ended topologies and the balanced topology is usually found in MMIC technology. In radio system applications, frequency doublers are usually designed for low power (< 10 mW), followed by a buffer amplifier. Although higher output power can be obtained from a multiplier, it is not a preferred approach due to the higher generation of unwanted harmonics.

One of the first reported works [1] on frequency multipliers employed a simple input matching circuit at the fundamental frequency, and an output matching circuit at the second harmonic, followed by a fundamental frequency trap. The preliminary reported result was 1 dB multiplication gain to double 4 to 8 GHz. This result originated a trend in the investigation of frequency multipliers using FETs that is still under way today. Later, higher output power and multiplication gain were reported, for instance, a frequency doubler from 4 to 8 GHz employing a 1×600 μm device presented a multiplication gain of 5 dB gain for an input power of 14 dBm [2]; the multiplication of 12 to 24 GHz using a device NE71082 by NEC provided 4 dB multiplication gain for an input power of 0 dBm [3].

A tentative way to provide good impedance match and control on the phase of reflected fundamental and second harmonic frequency signals [4] is described in Figure 7.1 for a doubler from 12 to 24 GHz. At the gate, a reflector (or trap) for the second harmonic was connected close to the device, while the fundamental frequency matching circuit was connected close to the generator. At the drain, a reflector for the fundamental frequency was also connected close to the device in cascade with the second harmonic matching circuit. The authors explored the dependence of multiplication gain on the electrical angles θ_1, θ_2. The characteristic impedance of all transmission lines

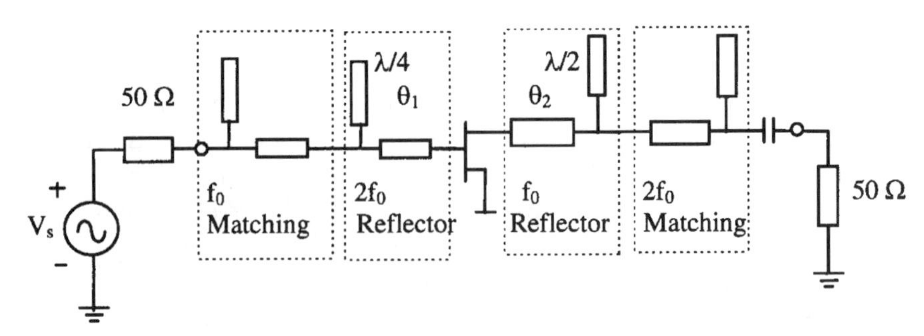

Figure 7.1 Frequency doubler with input/output reflector.

is equal to 50 Ω. The reported multiplication gain over the output frequency band of 23.2 to 24.2 GHz is 3 to 6 dB for an input power of +7 dBm.

An important topology when the output frequencies are in the millimeter-wave range is shown in Figure 7.2, where the input circuit is built in microstrip and the output is built in a waveguide, whose dimensions are practical to handle at these frequencies. The input circuit contains a blocking capacitor C_b, and matching transmission-line elements T_{L1}, T_{L2} and T_{L3}. The open-circuit stub T_{L3}, besides matching at the fundamental frequency, is also a filter for the second harmonic. Best efficiency is obtained by tuning the fundamental drain termination, adjusting the length of series transmission line, T_{L7}, from the drain to the microstrip-to-waveguide transition. Second harmonic impedance match is obtained by adjusting a tuning screw on the waveguide side. The transmission-line elements T_{L8} and T_{L9} are 90 degrees long at the fundamental frequency, and operate as a reject filter at that frequency. This filter is transparent to the second harmonic, and the other two elements, T_{L10} and T_{L11} whose dimensions are 45 degrees each, function as a second harmonic filter.

The waveguide is a high-pass filter and inherently rejects the fundamental frequency. Second and higher harmonics propagate through the waveguide. If the frequency doubler is properly designed, its RF current is an approximate half-wave sinusoid, where third harmonic content is low, and the fourth harmonic will be of much lower amplitude. A pioneer publication [5] in this area reported this topology, showing a multiplication gain of −1 dB at an output frequency of 30 GHz, and an output power of +8 dBm, employing a 0.5×250 μm device from Avantek.

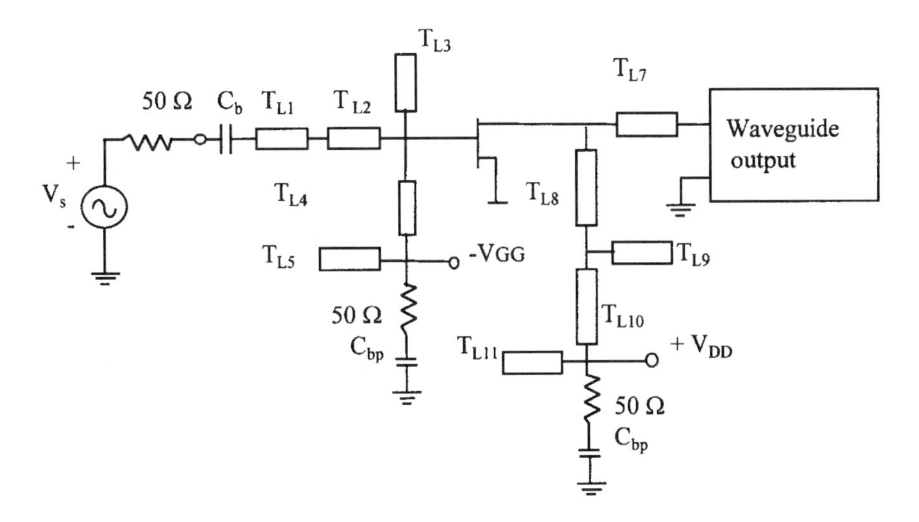

Figure 7.2 Frequency doubler employing waveguide output.

In all these designs a bandpass filter has to be coupled at the output to clean the spectrum. Another option to build a frequency doubler is to simply connect to the drain a bandpass filter centered at the second harmonic, in series with a transmission-line phase-shifter.

7.1.1 Tuned Frequency Doubler/Single-ended

The circuit shown in Figure 7.3 was designed to double the input frequency band tuned at 5 to 10 GHz. The gate contains the matching transmission-line elements T_{L1} and T_{L2}, and the bias filter elements T_{L3} and T_{L4}. The drain contains a transmission-line phase-shifter, T_{L5}, to adjust the phase of the fundamental frequency impedance and a harmonic filter which behave as a bandpass filter. It is composed of quarter-wavelength transmission lines, T_{L6}, T_{L7} and T_{L8}, which block the fundamental and third harmonic frequencies and present a 50 ohm termination at the second harmonic frequency. The drain bias filter is composed of elements T_{L9} and T_{L10} and their function is to isolate the bias from the generated second harmonic signals. The RC circuit, in parallel with the power supply, gives a resistive termination at low frequencies contributing to the circuit overall stabilization.

It has been demonstrated that biasing the device in the vicinity of pinch-off equates to best multiplication efficiency. This condition is obtained by applying a positive supply to the drain and a negative supply to the gate. A more practical application is to replace the negative supply with a self-bias circuit introduced by the source resistance, R_f, and a gate grounding resistor, R_B, indicated in the figure.

The calculation of R_f requires an initial estimate of dc drain current under RF driven conditions. Assuming maximum gate voltage swing from pinch-off to zero volts and a low drain impedance, the drain current will be

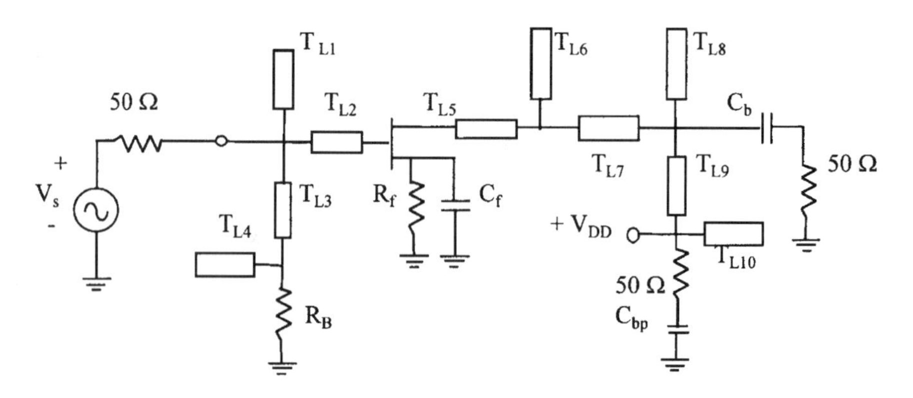

Figure 7.3 Tuned single ended frequency doubler.

rectified with I_{DSS} peak current. The dc component of a half-wave rectified sinusoid with a peak value of 60 mA is equal to $60/\pi \approx 20$ mA. The dc load line for the input circuit represented in Figure 7.4 is drawn from $V_{GS} = 0/I_{DS} = 0$ to $V_{GS} = V_P/I_{DS} = 20$ mA, which determines the correct value for R_f.

The quiescent voltage, $V_{G0} = -1.1$ V, and current, $I_{D0} = 15$ mA, is defined at the crossing of the input circuit dc load line and $I_{DS} - V_{GS}$ plot. Driving the device with a higher dynamic voltage, the current increases to 20 mA and the gate voltage rises to −1.5 volts. There is, however, a tradeoff on the value of R_f, due to the reduction of available dc voltage for the device. If this reduction can be compensated with a higher drain voltage then the performance of the multiplier with two supplies or one supply is approximately the same. The network elements obtained after computer optimization are given in Table 7.1.

The drain voltage and current waveforms obtained through simulation on MDS are shown in Figure 7.5. The drain current is measured at the device external terminal where the effects introduced by the device parasitics and wire bond connections are taken into account. The drain voltage is nearly sinusoidal, indicating a high rejection of fundamental and third harmonic components.

The signal trajectory on the I–V plane is shown in Figure 7.6, where one can observe the two distinct regions: the region where the device is on and the region where the device is off. The slope representing the "on" region corresponds to a load line equal to 40 ohms. This trajectory is very close to the one described in Figure 3.13 for a dc model.

The frequency response for this circuit in terms of multiplication gain and input return loss is shown in Figure 7.7 for the frequency range 4.5 to

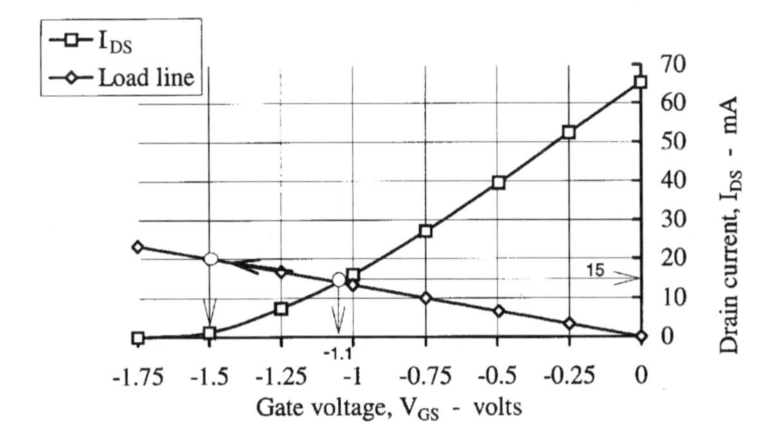

Figure 7.4 Determination of R_f.

Table 7.1
Frequency Doubler Network Elements (electric angles normalized at f_0 = 5.0 GHz)

Element No	Element Type	Parameters
T_{L1}	Open stub	Z_0 = 25 Ω/θ = 31.7°
T_{L2}	Series line	Z_0 = 80 Ω/θ = 65.4°
T_{L3}	Series line	Z_0 = 65 Ω/θ = 75.4°
T_{L4}	Open stub	Z_0 = 25 Ω/θ = 87.2°
T_{L5}	Series line	Z_0 = 50 Ω/θ = 25.7°
T_{L6}	Open stub	Z_0 = 50 Ω/θ = 79.25°
T_{L7}	Series line	Z_0 = 50 Ω/θ = 111.25°
T_{L8}	Open stub	Z_0 = 50 Ω/θ = 79.25°
T_{L9}	Series line	Z_0 = 90 Ω/θ = 45°
T_{L10}	Open stub	Z_0 = 25 Ω/θ = 45°
R_B	Resistor	R = 50 Ω
R_f	Resistor	R = 75 Ω
C_f	Capacitor	C = 20 pF

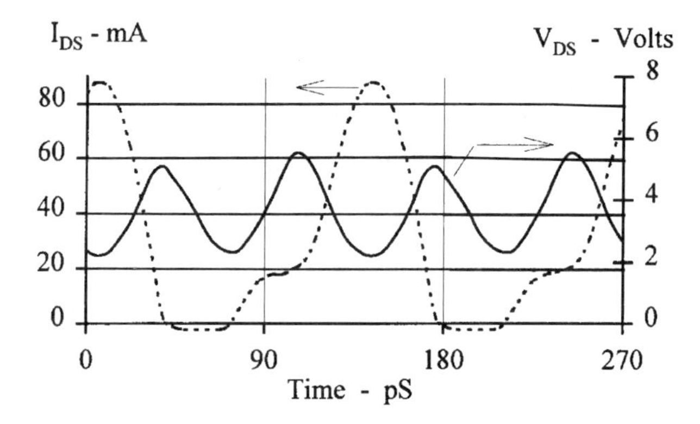

Figure 7.5 Drain current and voltage waveforms.

5.5 GHz. The maximum gain is on the order of 4.4 dB and the input return loss at the worst point is in the order of 7 dB at the low end of the band. This data was simulated for an input power of +10 dBm.

The effect of applied power in the multiplier performance is observed in the same plot, by changing the applied power to +5 dBm. Although the effect on the multiplication gain is negligible, it has a great impact on the input impedance. At lower drive levels, the multiplication gain drops and the return loss approaches 0 dB. At higher input power, the multiplication gain drops to a lower level corresponding to saturation of the multiplier, and the input

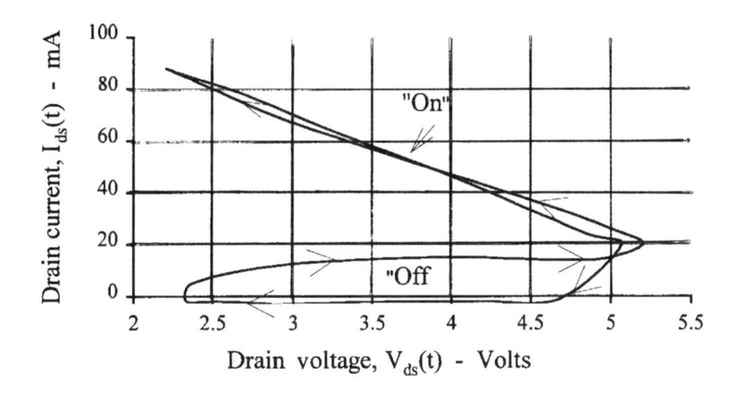

Figure 7.6 Signal trajectory on the I–V plane.

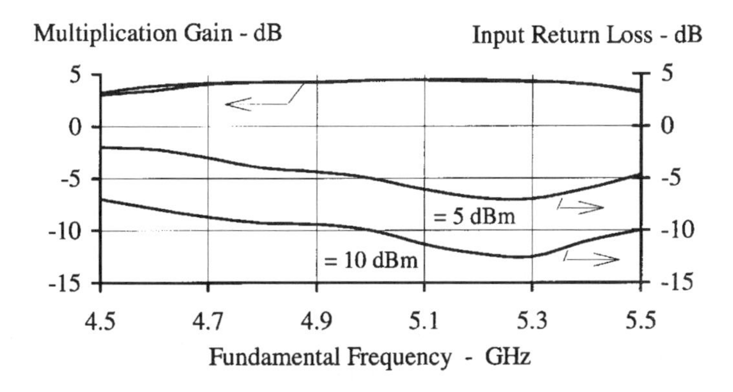

Figure 7.7 Frequency response of the tuned frequency doubler.

return loss improves. This performance is similar to those presented by class B amplifiers.

This is a medium-band design giving a reasonable performance over a 20% bandwidth. The bandwidth limiting factors are:

> *Drain side*: phase-shifter which resonates with the drain reactance at specific frequencies; effectiveness of open-circuited stubs in maintaining a short-circuit on the main output line over frequency.
> *Gate side*: high "Q" of the gate impedance.

Only the input return loss was simulated in this example and in the others in this chapter. As demonstrated in Chapter 5, simulation of output impedance for a linearized model is relatively straightforward. For the direct synthesis analysis it is also possible to obtain output impedance by repeating that process for each fundamental frequency. A special software would need to be developed for that purpose, otherwise this would be a tedious task. Most of the simulations in this chapter were carried out in MDS. It is not simple to obtain the output returns loss from this type of circuit using that software. It can, however, be obtained indirectly by making a load pull, i.e., changing the load impedance and observing the multiplier performance. That approach can be implemented in most software.

7.1.2 Wideband Frequency Doublers/Single-ended

The idea [6] of introducing frequency diplexers at the drain and gate of a frequency multiplier has the advantage of presenting resistive terminations over wide frequency range, minimizing signal reflections. One such configuration is shown in Figure 7.8, where the device is biased near pinch-off. The output diplexer is used to separate the second harmonic from the fundamental and other harmonic signals. It is composed of a bandpass filter (BPF), and a bandstop filter (BSF), for the second harmonic. The transmission line, T_{L5}, in series with the bandpass filter, adjusts the phase of the impedance at the fundamental frequency. The bandpass filter is of the edge coupled type containing at least three resonators, rejecting fundamental and third harmonic frequencies. The band stop filter composed of elements T_{L7}, T_{L8} and T_{L9}, blocks the second harmonic and presents low loss at fundamental frequency and third harmonic. The open stubs, T_{L8} and T_{L9} are a quarter-wave long at the second harmonic, and are spaced by a series line T_{L7} measuring 63.4 degrees at the fundamental frequency. Therefore, the bandstop filter presents a short-circuit to the second harmonic, and is matched to 50 ohms at the fundamental and third harmonic frequencies. Transmission line T_{L6} is a

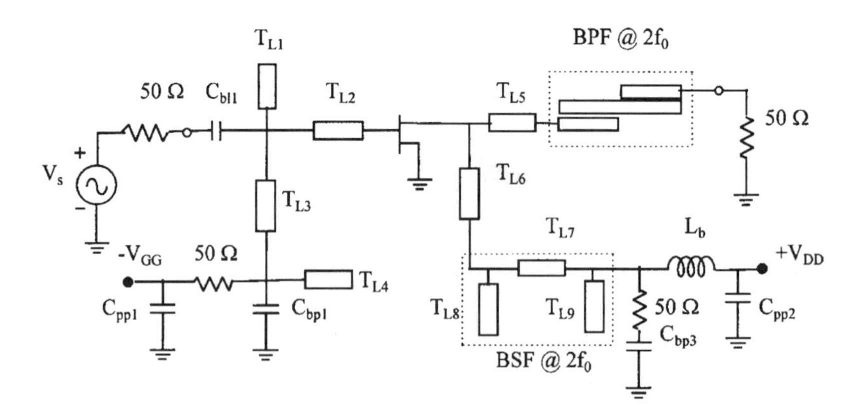

Figure 7.8 Wideband single ended frequency doubler.

quarterwave-long at the second harmonic transforming the short circuit to an open circuit at the drain.

An example of such an approach has been simulated for the input frequency band from 6 to 9 GHz, corresponding to 40% bandwidth. According to Figure 4.16, this wide band introduces a slope of 5 dB in the multiplication gain. Therefore, following wideband amplifier techniques, the circuit is ideally matched at the high end of the band, and mismatched at the low end to maintain the multiplication gain constant. Application of (4.32) provides the amount of mismatch at the low end of the band to maintain a constant multiplication gain, assuming a perfect match at the high end of the band.

$$|\Gamma_{in}| = \sqrt{1 - (1 - 3/9)2} = 0.745$$

Thus, a VSWR of 6.8 is expected at 6.0 GHz. A simulation of the doubler performance tuned at 9.0 GHz showed a multiplication gain of −2.6 dB, so that the overall multiplication gain within the band will be less than that value. The gate is matched at the fundamental frequency by a high impedance series transmission line, T_{L2}, and a low impedance open stub, T_{L1}, simulating a low pass circuit. Their parameters were optimized to obtain the objective of flat multiplication gain over the band. Harmonic signals generated by the device at the gate are absorbed by the input bias filter. This circuit presents excellent stability due to the drain resistive terminations at the fundamental and harmonic frequencies. The transmission-line phase-shifter was adjusted for maximum bandwidth. The elements obtained after optimization are described in Table 7.2.

This circuit is capable of providing 2.9 ± 0.6 dB loss when driven by a +10 dBm source sweeping from 6 to 9 GHz. The simulated performance for this multiplier is depicted in Figure 7.9 which shows multiplication gain and input return loss characteristics. Practical use of such a wideband multiplier would require either a 3 dB pad to raise the low input return loss at low frequencies to a reasonable 10 dB value, or a wideband isolator.

The harmonics dissipated in the bias load circuitry is depicted in Figure 7.10, where a high saturated power (+18 dBm) is at the fundamental frequency. The second harmonic bandstop filter also presents a narrow rejection band, from 7.2 to 8.4 GHz, dissipating nearly +5 mW power at the bandwidth extremes on the bias termination. If a better filter is applied, the multiplication efficiency can be improved at the bandwidth extremes.

7.1.3 Wideband Balanced Frequency Doubler

A means of improving the input impedance match in a wideband can be obtained by the use of a balanced topology [7], represented in Figure 7.11.

Table 7.2
Wideband Frequency Doubler Network Elements
(electric angles normalized at f_0 = 7.5 GHz)

Element No	Element Type	Parameters
T_{L1}	Open stub	$Z_0 = 40\ \Omega/\theta = 23.7°$
T_{L2}	Series line	$Z_0 = 75\ \Omega/\theta = 41.6°$
T_{L3}	Series line	$Z_0 = 70\ \Omega/\theta = 95.0°$
T_{L4}	Open stub	$Z_0 = 20\ \Omega/\theta = 130.9°$
T_{L5}	Series line	$Z_0 = 50\ \Omega/\theta = 53.4°$
T_{L6}	Series line	$Z_0 = 50\ \Omega/\theta = 38.6°$
T_{L7}	Series line	$Z_0 = 50\ \Omega/\theta = 35.7°$
T_{L8}	Open stub	$Z_0 = 50\ \Omega/\theta = 45°$
T_{L9}	Open stub	$Z_0 = 50\ \Omega/\theta = 45°$
R_B	Resistor	$R = 50\ \Omega$
$C_{bp1} = C_{bp3}$	Capacitor	C = 20 pF
$C_{pp1} = C_{pp2}$	Capacitor	C = 1000 pF

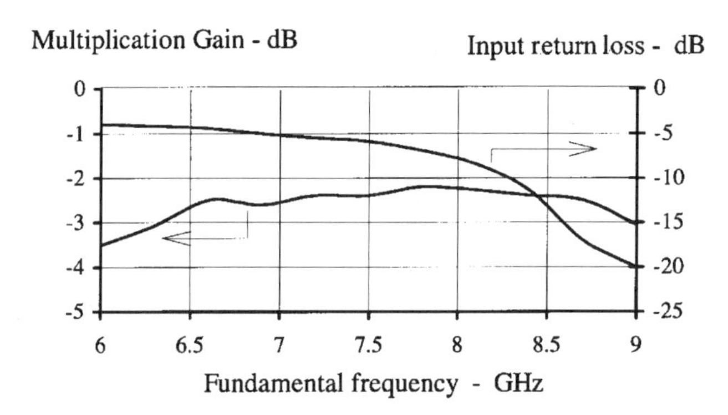

Figure 7.9 Wideband frequency doubler performance.

Inserting two FET half-wave rectifiers antisymmetrically so that conduction occurs on alternating half-wave cycles, one can obtain an active full-wave rectifier. The input coupler and output couplers must present a bandwidth wide enough to cover the fundamental and the second harmonic, which can be achieved by Lange couplers.

Besides the VSWR advantage at both the input and output ports, this topology presents good isolation between the multiplier stages, and is very stable since the devices are terminated into 50 ohm over a wideband frequency. The input coupler introduces a 90-degree phase shift and the output coupler another 90 degrees, so that the current from each device at the output load

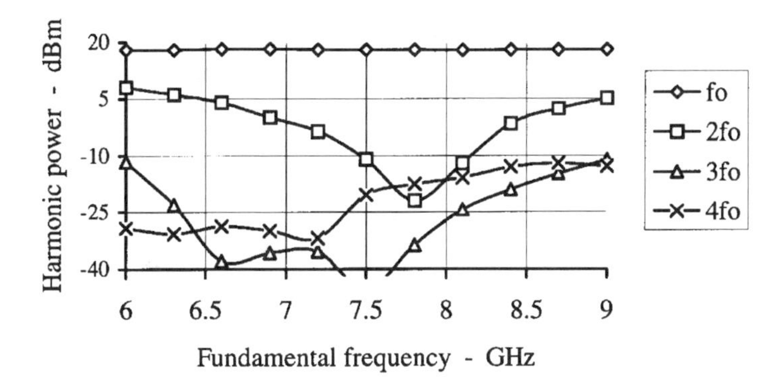

Figure 7.10 Harmonic power on the bias termination.

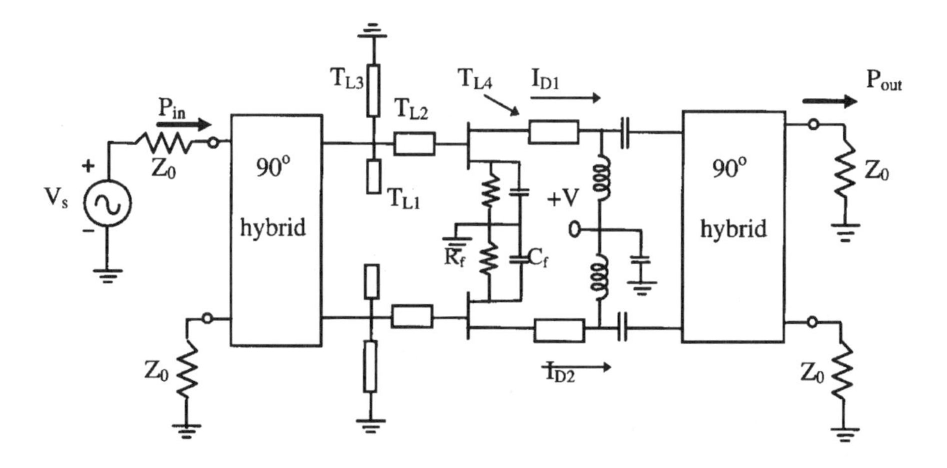

Figure 7.11 Wideband balanced frequency doubler.

are given by (7.1) and (7.2). Therefore, the odd harmonics are 180 degrees out of phase, are canceled at the output port, and dissipated at the coupler termination. The even harmonics at the output port are in phase and added in power.

$$I_{D1} = \frac{I_F}{\pi} + \frac{I_F}{2}\cos(\omega t) + \frac{2I_F}{3\pi}\cos(2\omega t) + \ldots \tag{7.1}$$

$$I_{D2} = \frac{I_F}{\pi} + \frac{I_F}{2}\cos(\omega t + \pi) + \frac{2I_F}{3\pi}\cos(2\omega t + 2\pi) + \ldots \tag{7.2}$$

In practical applications, the input matching circuit is designed for a wide band at the fundamental frequency, but may introduce a high impedance mismatch at the second harmonic frequency band. If that happens, the input coupler may not be effective at the second harmonic. The gate matching circuit in this example is the same used in the previous example. It was modified to incorporate the self-bias circuit, consisting in the grounding of transmission line T_{L3} and inserting the bias resistor, R_f, and bypass capacitor C_f. The bias is adjusted such that the gate voltage is near pinch-off when the device is driven by the RF signal. The bandwidth limitation of this configuration is mainly due to the input matching. The output impedance is not matched to avoid introducing bandwidth restrictions.

A set of simulations were carried out for a bias voltage of +5V and an input power of +10 dBm. The results showed this topology can deliver a multiplication gain of −5.0 dB, flat within 0.5 dB, for an input frequency from 6 to 9 GHz. The difference in gain compared to the previous topology is on the lack of a drain element to resonate the device output reactance at the fundamental frequency. The transmission-line element connects the drain to the coupler impedance, 50 ohms, instead of connecting to a reactive termination. As observed in Figure 7.12, the input matching is on the order of 20 dB across the band. These results consider both devices identical in their dc and RF characteristics. If they are not equal, the input VSWR will degrade proportionally. This is a good topology for MMIC design, where similar devices are readily available.

The simulation of drain currents for each device is shown in Figure 7.13, where a 90 degree phase shift between each waveform is seen. The distortion is antisymmetric, indicative of a waveform rich in even harmonics. The effect of the 50 ohm drain loading at the fundamental frequency is observed in the maximum peak current that is lower than I_F.

An additional simulation was carried with this topology to find out its performance as a function of drive level. The result of Figure 7.14 shows a

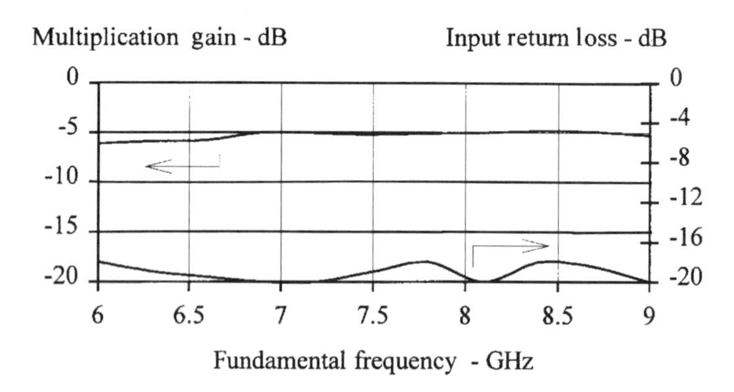

Figure 7.12 Frequency response for balanced frequency doubler.

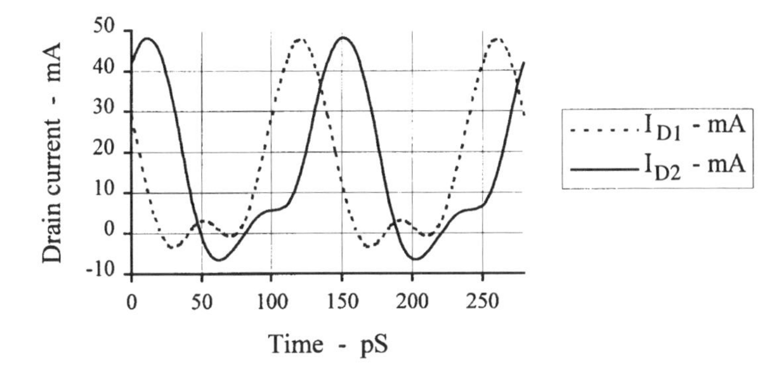

Figure 7.13 Drain current waveforms for balanced configuration.

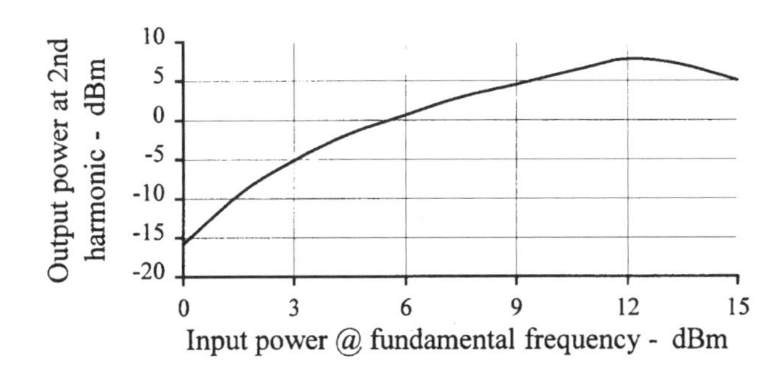

Figure 7.14 Output/input power relationship for a balanced frequency doubler. Fundamental frequency = 7.5 GHz.

monotonic relation from 0 to +15 dBm input power. Maximum output power obtained is on the order of +8 dBm, with +12 dBm of input drive corresponding to −4 dB multiplication gain. This simulation provided important information related to the range of drive level that can be applied to such a device. For instance, the input power variation from 9 to 15 dBm is reduced to 2.5 dB at the output, contributing to better stability over temperature. However, a maximum drive power has to be determined to avoid introducing excess phase noise into the original signal.

7.1.4 Balanced/Unbalanced Frequency Doubler

Another approach [8] to balance a frequency doubler is to use an input 180 degree hybrid to drive each device in antiphase. Therefore, the fundamental drain currents are also in antiphase and a good rejection is obtained by paralleling both drains. As a matter of fact, the connection point becomes a virtual ground and the length of line from the junction of both transmission lines to the drain becomes an inductive short stub that can be used for tuning the drain at the fundamental frequency. It is observed by the set of (7.1) and (7.2), which are also valid for this configuration, that the resulting second harmonic from each device are in phase, and are therefore added in the output load.

The bandwidth limitation becomes a function of the transformer "balun" used, the gate Q-factor, and of the drain tuning line. An example of a microstrip version, using a ratrace coupler as the 180 degree hybrid is depicted in Figure 7.15 for operation within the fundamental bandwidth, 10 to 12 GHz. Calculating the maximum multiplication gain at the band extremes, it was found 0.0 dB at the low end and −3.0 dB at the high end. Therefore, the matching circuit is designed to obtain maximum gain at 12.0 GHz and mismatch the impedance at the low end to maintain a constant multiplication gain. The input circuit is symmetrical and comprises an open stub, T_{L6}, and a series transmission line, T_{L7}. At the drain, both devices are connected together to the same output line. The drain is biased by the inductance L_{CK}. The ratrace coupler is composed of transmission-line elements T_{L1}, T_{L4} (70.7Ω/90°) and T_{L5} (70.7Ω/180°).

This type of balun is simple to construct and has a termination for absorbing any in-phase voltages, helping to stabilize the circuit. The simulation for this frequency doubler provided an average −5 dB multiplication gain over the input frequency band from 10 to 12 GHz, as shown in Figure 7.16. Note the device is capable of generating frequencies as high as 24 GHz, which is above the frequency of operation for this device as an amplifier.

The disadvantage of this topology is its higher susceptibility to imbalances in device dc parameters and RF matching networks compared with the previous

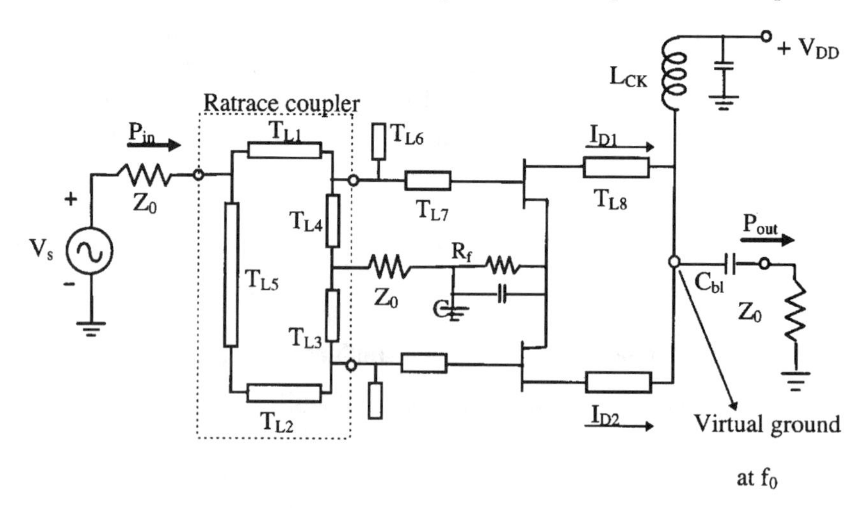

Figure 7.15 Balanced/unbalanced frequency doubler.

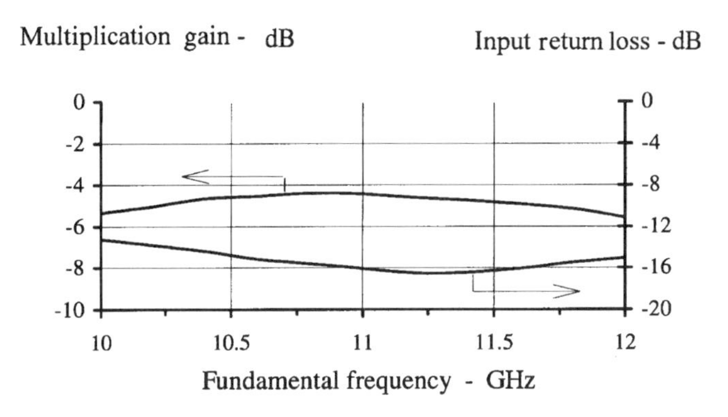

Figure 7.16 Performance of balanced/unbalanced frequency doubler.

topology. Any imbalance is reflected back to the generator, therefore requiring an attenuator at the input to minimize the resulting standing wave.

An example of a frequency doubler employing this topology is depicted in the photo of Figure 7.17, which shows the realization on 10 mils thick alumina substrate of a frequency doubler operating from 6.0 to 9.0 GHz at the fundamental frequency. Observe that the traditional ratrace coupler has been modified to operate over the desired band. This circuit employs the

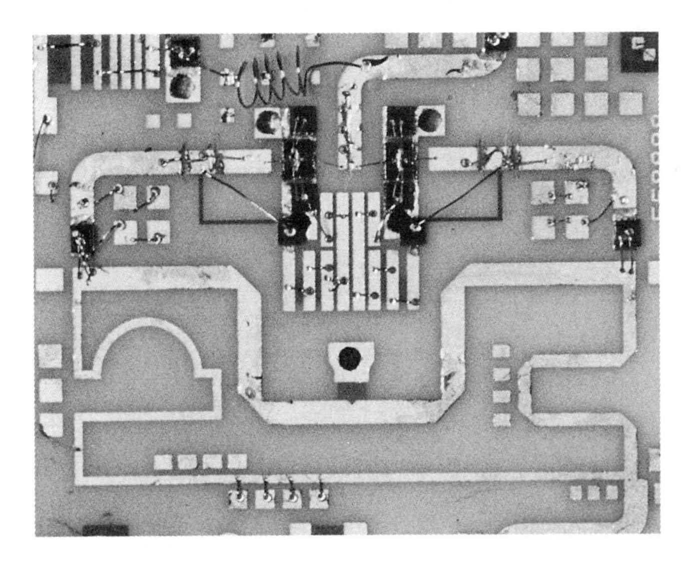

Figure 7.17 Wideband balanced/unbalaned frequency doubler operating from 6.0 to 9.0 GHz (*Courtesy of Hewlett-Packard Company, Wireless Infrastructure Division*).

reference device and provided a multiplication gain of −2.0 ± 0.25 dB over the band.

7.2 Frequency Triplers

The subject of FET frequency triplers took longer to be reported compared to frequency doublers, due to the special operating conditions required to obtain a reasonable performance. A simple drain circuit using a bandpass filter and an open stub to adjust optimum fundamental and second harmonic termination, was reported [3] to present 3 dB multiplication gain from 4 to 12 GHz with a drive of +10 dBm. This result was obtained with a 1 × 600 μm device biased near zero gate volts. Published results on a topology [6], similar to the one presented in Figure 7.18, presented a conversion loss of 10.6 ± 0.6 dB over the fundamental frequency band 8.5 to 10.5 GHz, with an input drive of 10 mW.

7.2.1 Frequency Tripler/Single-ended

A topology similar to that of Figure 7.8 was applied to a frequency tripler, with a different bias, impedance matching and harmonic terminations. It was demonstrated in Chapter 3 that the adequate bias and drain termination for odd harmonic generation should emphasize voltage distortion instead of current. Therefore, the phase-shifter in series with the bandpass filter has to provide a parallel resonance of the drain parasitics in order to maximize the voltage on the device nonlinear elements at the fundamental frequency. The active device is then biased in class A, and a large input power is applied to swing the gate

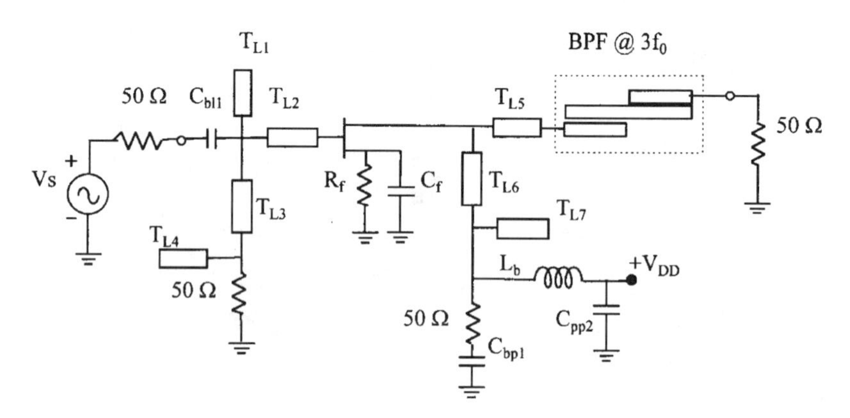

Figure 7.18 Single-ended frequency tripler.

from pinch-off to zero gate voltage. The circuit topology shown in Figure 7.18 was designed to multiply by three the fundamental frequency band from 5.5 to 17.5 GHz. The circuit elements are described in Table 7.3. The device is biased class A (V_{DD} = 3.5 volts; I_{D0} = 32 mA) and driven by a +10 dBm signal.

The drain circuit is composed by a bandpass filter for delivering the third harmonic to the load, in series with the transmission-line filter T_{L5}. The bias is delivered through the transmission-line elements, T_{L6} and T_{L7}, that block the third harmonic frequency, and present a low pass type of response to the fundamental and second harmonic.

The purpose of the input bias filter is similar to the one used in the frequency doubler: couple the generator to the gate and absorb all reflected harmonics. The bandwidth of such a multiplier is essentially limited by the drain phase-shifter's ability to resonate the device output parasitics, and the gate impedance mismatch with frequency. The input return loss of such a circuit will be dependent on the device input impedance and may require an input isolator for proper operation. A simulation of the drain voltage for the frequency tripler is illustrated in Figure 7.19.

In the figure, distortion is on both sides of the voltage waveform, an indication of the high content of odd harmonics. Also observed are high voltage peaks, indicative of a high impedance termination. The drain current waveform is shown in Figure 7.20, and a third harmonic sinusoidal was inserted into the plot, represented by a dotted line to emphasize the third harmonic content. The drain current peaks at a value higher than I_{DSS} with gate conduction for a brief period of the microwave signal.

Table 7.3
Single-ended Frequency Tripler Network Elements
(electric angles normalized at f_0 = 5.5 GHz)

Element No	Element Type	Parameters
T_{L1}	Open stub	$Z_0 = 35\ \Omega/\theta = 8.7°$
T_{L2}	Series line	$Z_0 = 82\ \Omega/\theta = 17.4°$
T_{L3}	Series line	$Z_0 = 50\ \Omega/\theta = 58.0°$
T_{L4}	Open stub	$Z_0 = 25\ \Omega/\theta = 95.9°$
T_{L5}	Series line	$Z_0 = 50\ \Omega/\theta = 17.4°$
T_{L6}	Series line	$Z_0 = 80\ \Omega/\theta = 30.0°$
T_{L7}	Open stub	$Z_0 = 30\ \Omega/\theta = 30.0°$
R_f	Resistor	$R = 20\ \Omega$
$C_{bp1} = C_f$	Capacitor	$C = C_f = 20$ pF
C_{pp1}	Capacitor	$C = 1000$ pF

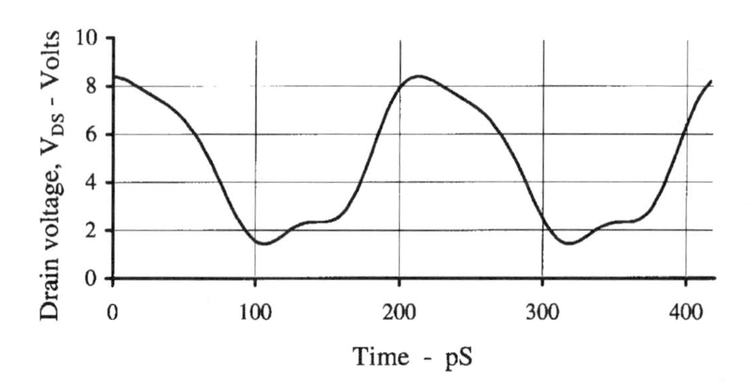

Figure 7.19 Drain voltage waveform for single-ended frequency tripler.

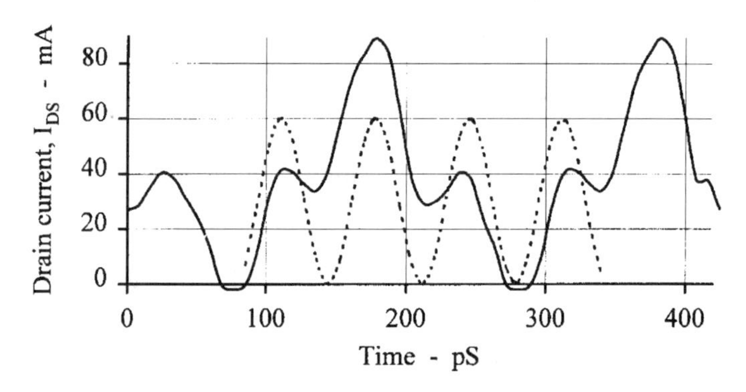

Figure 7.20 Drain current waveform for a single-ended frequency tripler.

An interesting signal trajectory can be observed on the I–V plot, which is seen in Figure 7.21, for this tripler and is similar to the one described in Figure 3.17 based on a dc model. Two distinct regions are depicted, one near pinch-off where drain voltage exceeds 8V, and the other near the drain saturation, where the drain current exceeds 80 mA, suggesting gate diode conduction. The transition between both these regions follows a load line of 60 ohms. The quiescent bias point is depicted by point A on the figure.

The performance of this circuit for a +10 dBm drive from 5 to 6 GHz is shown in Figure 7.22. A multiplication gain in the order of −4 dB is observed with an input return loss close to 10 dB. In spite of the negative gain presented with this circuit, its losses are lower than the losses obtained with an equivalent diode tripler [9], the latter being as high as 16 dB.

7.2.2 Push-pull Frequency Tripler

The "push-pull" amplifier, represented in Figure 7.23, consists of two amplifiers inserted within two 180 hybrid couplers, or "baluns." This topology is exten-

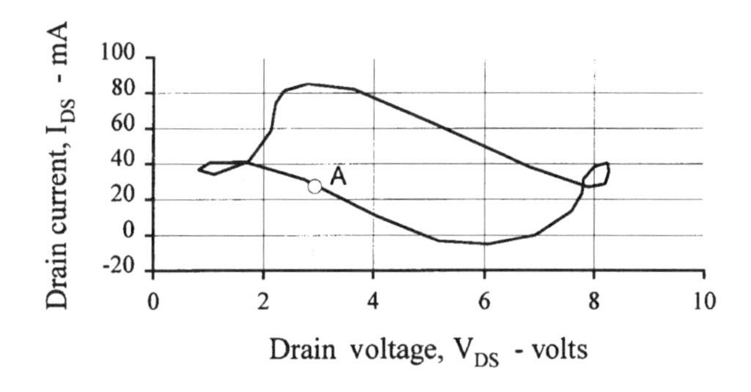

Figure 7.21 Trajectory for a single ended frequency tripler.

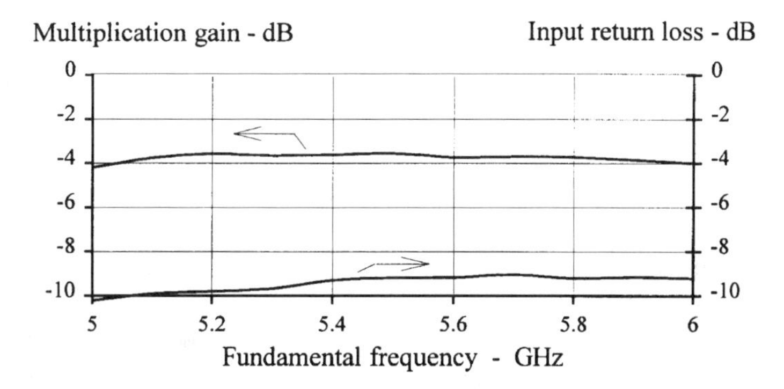

Figure 7.22 Single-ended frequency tripler performance.

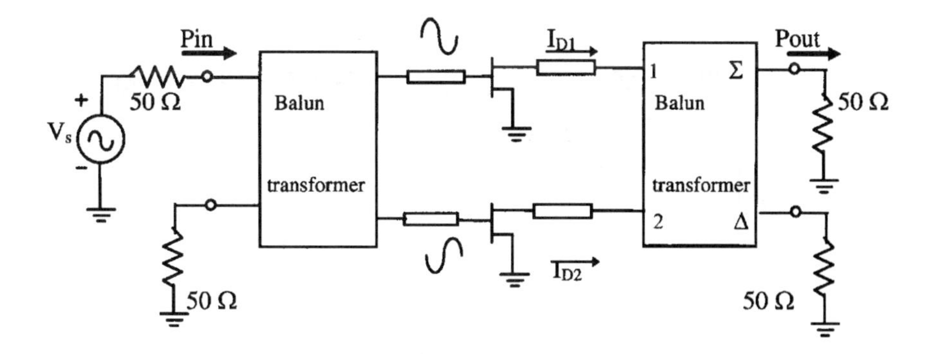

Figure 7.23 Push-pull amplifier.

sively used in the design of wideband power amplifiers due to its properties of cancellation of harmonics and phase addition of the fundamental frequency component. If wideband transformers are employed, then wide band amplifiers are obtained with very low even harmonic content. The demonstration of this latter fact is shown by (7.3) and (7.4), which describe the harmonic currents at the drain of two FETs driven by a 180 degree transformer.

$$I_{D1} = I_{d0} + I_{d1}\cos(\omega t) + I_{d2}\cos(2\omega t) + I_{d3}\cos(3\omega t) \ldots \quad (7.3)$$

$$I_{D2} = I_{d0} + I_{d1}\cos(\omega t + \pi) + I_{d2}\cos(2\omega t + 2\pi) + I_{d3}\cos(3\omega t + 3\pi) \ldots \quad (7.4)$$

The output transformer introduces additional $n180$ degree (n = harmonic order) rotation on current I_{D2}, so that at the load, only the fundamental components of both currents are in phase. All others are in antiphase and are ideally canceled.

In the case of a balanced frequency tripler, the advantage of push-pull effect can be taken by first observing that the drain currents are balanced at the fundamental frequency and at the third harmonic frequency. Then, by employing an output transformer presenting a transfer characteristic from port 1 to the output port, Σ, common to that from port 2 to output port at the fundamental frequency, and a differential transfer characteristic from the same ports at the third harmonic. In these conditions, the fundamental frequency currents are canceled and the third harmonic currents are added in phase. For instance, the plot of Figure 7.24 shows the performance of a ratrace hybrid designed for 15 GHz, where the power is combined for signals with same amplitude and with a phase difference of 180 degrees. One can observe that

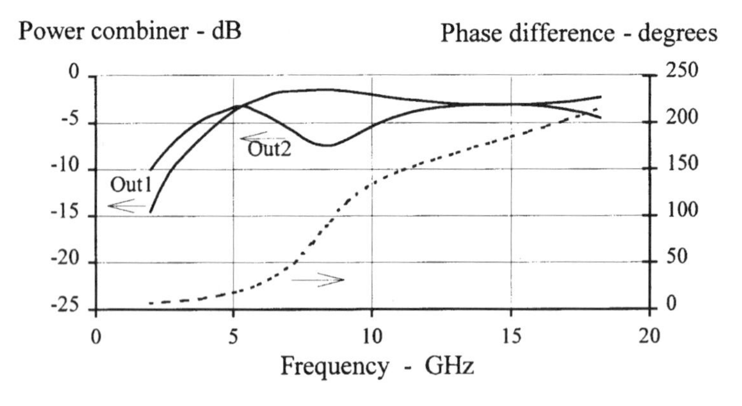

Figure 7.24 Performance of a ratrace hybrid.

at 5 GHz,the amplitudes and the phases are nearly equal, so that the differential drain mode impedance is close to a short circuit. With such a component, the fundamental frequency is attenuated and the third harmonic is added in-phase at the load.

Therefore, applying a ratrace at the input centered at the fundamental frequency, and another at the output centered at the third harmonic, one is able to take advantage of similar push-pull properties. An example of such a topology is presented in Figure 7.25.

The input match contains a series inductance simulated by a high impedance transmission-line, T_{L2}, and a parallel capacitor simulated by an open-circuited stub, T_{L1}. Connecting both inputs, there is a quarter-wave long transmission line, T_{L3}, that introduces a short at the second harmonic, improving the performance. The output match contains only a quarter-wave high impedance transmission line, T_{L4}, to parallel tune the drain output impedance at the fundamental frequency. The devices are self-biased and the source resistor does not need to be bypassed due to the phases of the two currents at each source which are in antiphase. In practice, bypass capacitors need to be added because any asymmetry in the devices unbalances the currents and then power is dissipated in the source resistors.

The network elements are described in Table 7.4 for one side of the circuit. The other side is symmetrical. The ratrace circuit used is conventional, with 70.7 ohms line impedance. The input coupler is tuned at 5.0 GHz and the output one to 15 GHz.

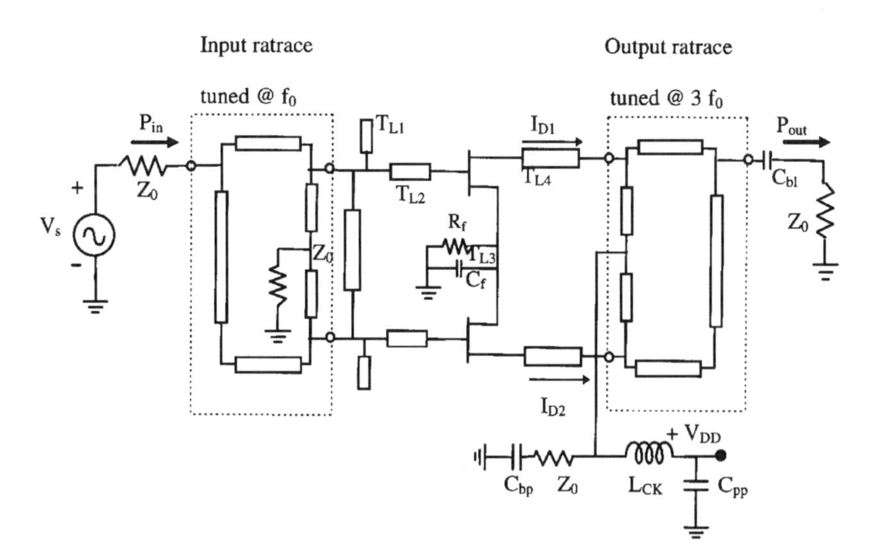

Figure 7.25 Push-pull frequency tripler.

Table 7.4
Push-Pull Frequency Tripler Network Elements
(electric angles normalized at f_0 = 5.0 GHz)

Element No	Element Type	Parameters
T_{L1}	Open stub	$Z_0 = 50\ \Omega/\theta = 15.8°$
T_{L2}	Series line	$Z_0 = 30\ \Omega/\theta = 19.8°$
T_{L3}	Series line	$Z_0 = 70\ \Omega/\theta = 80.8°$
T_{L4}	Series line	$Z_0 = 50\ \Omega/\theta = 47.6°$

The devices are self-biased with a 20 ohm resistor, resulting in a quiescent current of 30 mA, at V_{DD} = 4.0V. The multiplication gain for a 10 dBm input drive shown in Figure 7.26 is better than −3 dB, and the input return loss at the worst point in the band is −8 dB.

The drain current waveforms for both drains are shown in Figure 7.27, where both currents are not symmetrical, due to the unequal impedances presented by the ratrace at the fundamental frequency.

This multiplier gain compression is depicted in Figure 7.28 at 5 GHz. A linear multiplication gain is observed at low level, and 1 dB compression is at an input power of 6 dBm. At higher drive levels, the multiplier enters smoothly into saturation. At an input power, 10 dBm, the output power is in the order of +6.5 dBm.

7.3 Frequency Quadruplers

Frequency quadruplers are less efficient than triplers and doublers, but its design simplicity compared to diode multipliers is gaining acceptance within

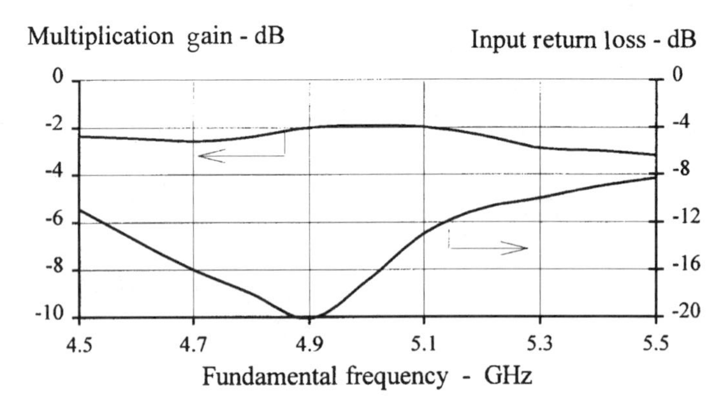

Figure 7.26 Performance of push-pull frequency tripler.

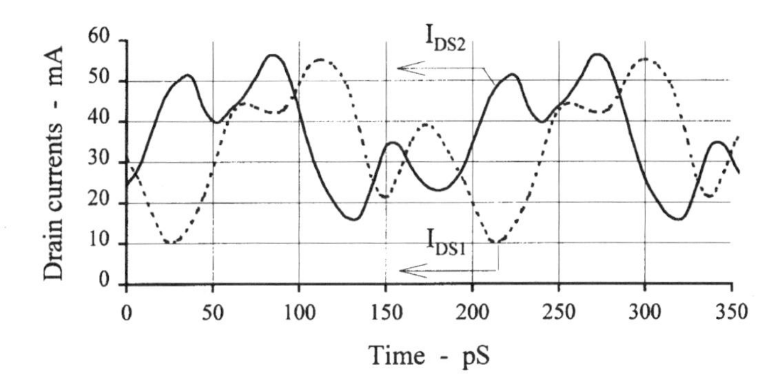

Figure 7.27 Drain currents for the push-pull frequency tripler.

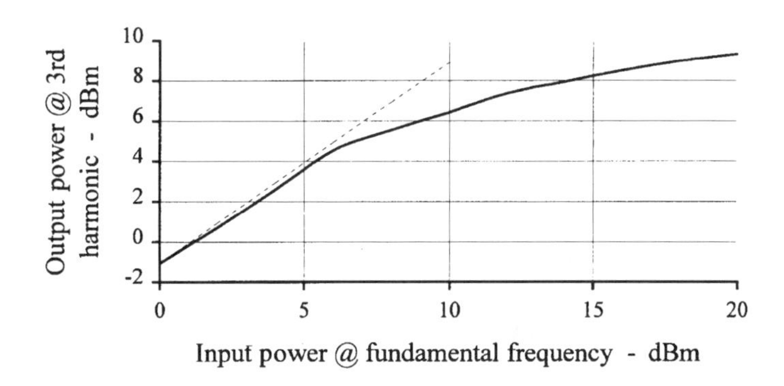

Figure 7.28 Compression characteristic of the push-pull frequency tripler.

the industry. Previous reported results using a $1 \times 600\ \mu m$ for a 4 to 16 GHz circuit using a topology similar to one described in 7.18 provided a maximum gain of 2 dB with the device biased in class B. A similar result on a frequency quadrupler was reported [10], which used a device type NE 71083 by NEC, possessing a gate area of $0.3 \times 380\ \mu m^2$. Although bandwidth is not described, a conversion loss of 6.5 dB was reported for a fundamental frequency of 5 GHz.

A frequency quadrupler can be obtained as an extension of a frequency doubler, where the harmonic filter is modified to reject fundamental, second and third harmonics. According to Table 3.1, the magnitude of the generated 4th harmonic is only 3 dB lower than the second and the third harmonic if the conduction angle is 60 degrees. However, that requires biasing the device in deep class C operation, where a high input power has to be applied to the device and the input impedance is difficult to match. If class B operation is used, then the 4th harmonic is 15 dB lower than the second. One way to partially overcome this problem is to use a larger gate device, so that with higher current and consequently higher transconductance, one can build a multiplier presenting a reasonable performance. For this exercise, the reference device had its gate area doubled. In consequence, I_{DSS} is doubled, as well as the device capacitances. A conventional topology for this type of multiplier is similar to the one illustrated in Figure 7.3. The main difference is on the second open-circuited stub connected to the drain that is made half the size of the first to reject the second harmonic and present low loss to the fourth harmonic. The element values reference to that circuit are described in Table 7.5.

The frequency response for this circuit in terms of multiplication gain and return loss are shown in Figure 7.29 for the fundamental frequency range 4.5 to 5.5 GHz. The drive power for this circuit is +10 dBm, and the device was biased at V_{DD} = 5V, I_{D0} = 10 mA. A multiplication gain of −7 dB was obtained from 4.7 to 5.3 GHz, with an input return loss better than 8 dB. Observe from this plot that the bandwidth is in the order of 10%, representing a trade-off with input match, which was kept in the order of 8 dB. Better input match and efficiency can be obtained at the expense of bandwidth. The harmonic filter is adequate to provide the required terminations to the drain. However, it does not suffice to produce a clean output spectrum. A bandpass filter must therefore be added at the output.

The simulated drain current and voltage are in Figure 7.30, a distorted drain current, as well as a distorted drain voltage. Both waveforms are contribut-

Table 7.5
Frequency Quadrupler Network Elements
(electric angles normalized at f_0 = 5.0 GHz)

Element No	Element Type	Parameters
T_{L1}	Open stub	$Z_0 = 35\ \Omega/\theta = 7.9°$
T_{L2}	Series line	$Z_0 = 75\ \Omega/\theta = 15.2°$
T_{L3}	Series line	$Z_0 = 65\ \Omega/\theta = 15.8°$
T_{L4}	Open stub	$Z_0 = 25\ \Omega/\theta = 87.2°$
T_{L5}	Series line	$Z_0 = 60\ \Omega/\theta = 25.7°$
T_{L6}	Open stub	$Z_0 = 40\ \Omega/\theta = 87.2°$
T_{L7}	Series line	$Z_0 = 50\ \Omega/\theta = 33.7°$
T_{L8}	Open stub	$Z_0 = 50\ \Omega/\theta = 43.5°$
T_{L9}	Series line	$Z_0 = 90\ \Omega/\theta = 17.8°$
T_{L10}	Open stub	$Z_0 = 35\ \Omega/\theta = 39.6°$
R_B	Resistor	$R = 50\ \Omega$
R_f	Resistor	$R = 75\ \Omega$
C_f	Capacitor	$C = 20$ pF

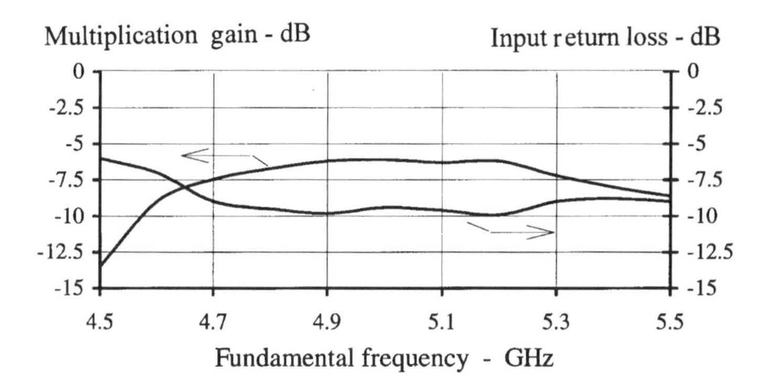

Figure 7.29 Frequency response of the tuned frequency quadrupler.

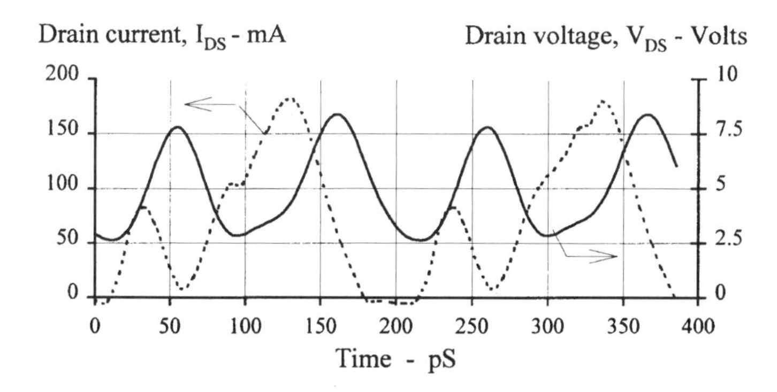

Figure 7.30 Drain current and voltage as a function of time.

ing to the 4th harmonic generation, or both the transconductance and output conductance are the important nonlinear generators.

The resulting signal trajectory on the I–V plane is represented in Figure 7.31, where the wider loop encompass the smaller one. In one fourth of the cycle, the device is completely pinched off, while it is at maximum voltage and current in another quarter of a cycle. This particular circuit does not reach drain voltage saturation at this power level.

7.4 Higher Order Frequency Multipliers

The direct generation of harmonics of higher order can be obtained by biasing the device at different conduction angles, as indicated in Table 3.1. However, they all require biasing the device in class C, which is more difficult to match at high frequencies.The bias alternatives are the order of dependent on a multiplier.

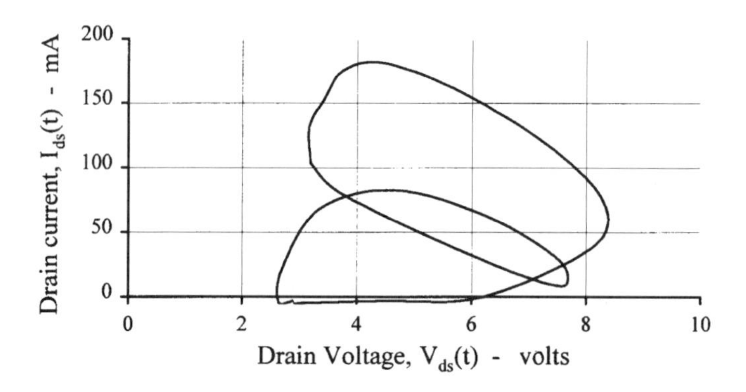

Figure 7.31 Signal trajectory for a frequency quadrupler.

Even order

In this case, the device is biased class AB similar to that employed in a frequency doubler, in order to obtain a rectified sinusoidal drain current which is rich in even order harmonics. A rectified drain voltage waveform is also used, where the device is biased close to zero gate bias. In the latter case, the harmonic load has to be able to short-circuit the appropriate harmonics, in order to maintain the desired output distorted waveform.

Odd order

The optimum bias condition for this class of multiplier is the one that generates an output waveform with distorted positive and negative peaks. The first option would be to bias the device in class A, about the center of the drain current *x* gate voltage transfer characteristic and apply a high power at the gate.

The magnitude of higher harmonic components like 5th, 6th, 7th, etc, becomes too small, requiring a high load resistance to compensate the reduction in output power. However, that procedure would apply as long as the load is smaller or equal than the device drain-to-source resistance. Even though higher-order harmonics are of low amplitude, the generation efficiency may be improved by taking advantage of mixing effects on the device nonlinearities if the device terminations are properly phased as previously discussed. The output harmonic is then matched to its optimum impedance determined by the harmonic current component and the drain voltage swing. The FET equivalent circuit can again be broken into its Fourier current source components, allowing the determination of input impedance and drain impedance.

Published [11] results for a x5 multiplier using the NE 67483 device from NEC, showed a multiplication gain of −7 dB from 2 to 10 GHz, for an applied input power of +12 dBm. The multiplier was built using open-circuited stubs to reject 2nd, 4th and 6th harmonic, followed by a bandpass filter that cleaned the output spectrum.

7.5 MMIC Multipliers

In recent years there has been considerable development of MMIC technology, resulting in availability in the market of complete functions in a single die at competitive cost compared with discrete versions. A revolution similar to the one introduced by the Silicon ICs in the early seventies is taking place with GaAs Microwave ICs which is changing the way subassemblies are built. In this context, MMIC multipliers are divided into two categories, one where multiplier chips are specifically designed for this technology and one using commercially available amplifier chips.

7.5.1 Custom MMICs

This is the most adequate approach from the point of view of electrical performance and overall efficiency, but the trade-off is cost. This is an expensive technology, and such an approach is adequate if a reasonable volume is envisioned. A low cost MMIC uses minimum GaAs area, so lumped elements or even active elements are preferred rather than transmission lines. Any of the multiplier topologies presented so far can be built in MMIC form. The single-ended topology of Figure 7.3 has been used recently [12] on a 30 to 60 GHz doubler. The circuit was built using High Electron Mobility Transistor (HEMT) technology and showed a multiplication gain of −1.5 dB at an input power of 7 dBm with a 10% bandwidth centered at 30 GHz. As pointed out previously, MMIC technology is also adequate to build balanced multipliers where nearly identical devices can be used. An example of application of balanced/unbalanced topology described in Figure 7.17 has been described [13] in the design of a 13 to 26 GHz doubler utilizing MESFET MMIC technology. The first problem to design such a circuit is to define how to build a miniature balun. Conventional topologies using a microstrip balun can be used at millimeter wave frequencies where the physical dimensions are acceptable to be inserted on the chip, but at this frequency an alternative was proposed. The authors used a coplanar to slotline transition, and then each side of the slot was connected to the gate. The circuit presented less than –4 dB multiplication gain centered at 12.25 GHz, and a bandwidth slightly higher than 10% was obtained. The maximum output power obtained was +6 dBm when biased at 3.0V/25 mA; the rejection of the fundamental frequency is greater than 35 dB. This type of performance simplifies post-filtering and indicates how good the circuit is balanced.

One application of active baluns has been reported [8], where the signal applied to the gate is extracted from the drain and source simultaneously, shown in Figure 7.32. At low frequencies, a signal applied to the gate of the active balun will generate a drain current proportional to the magnitude and phase of the applied voltage. The drain current will develop voltages on resistors R_{s1} and R_{d1}. In the former resistor, the source voltage will be in phase to the gate voltage due to the direct increase in current. In the latter, the increase in drain current will make the drain to ground voltage decrease due to the voltage drop on that resistor. Therefore, the drain voltage and source voltage will be 180 degrees out of phase and the magnitude can be made equal by the right choice of resistors. At high frequencies, different C_{gs} and C_{gd} capacitances generate different phase shifts at each terminal. They can, however, be compensated by adding a short transmission-line in the source terminal. Power will be dissipated in those resistors, which is a drawback compared to other alternatives. However, it is small and adequate for MMIC applications. The doubling action is performed by the other two transistors, where the input matching is carried out by transmission-lines T_{L4} and T_{L5}, and the output reactance is resonated by the drain series transmission-line, T_{L6}. The bias circuit for the multiplier devices in this example is done off chip.

A miniaturized MMIC doubler topology has been proposed [14], where common source and common gate devices are used to create the 180 degree phase shift, illustrated in Figure 7.33(a). The gate of device 1 is grounded by capacitor C1, so that the source to ground voltage is applied reversely to the gate-to-source. Thus, an increase in the signal voltage applied to the source will increase the gate-source voltage in the negative sense with a consequent decrease in the drain-to-source current. The drain potential will increase assuming it is connected to a positive supply by means of a resistance. Thus, the phase of applied voltage on the source will result in the same phase of the

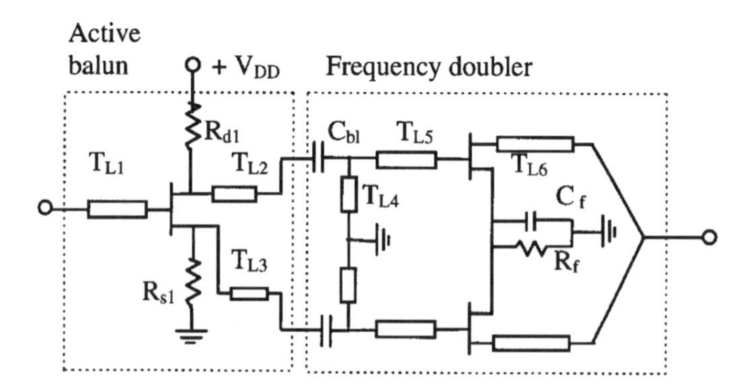

Figure 7.32 Monolithic version of 180-degree balanced frequency doubler.

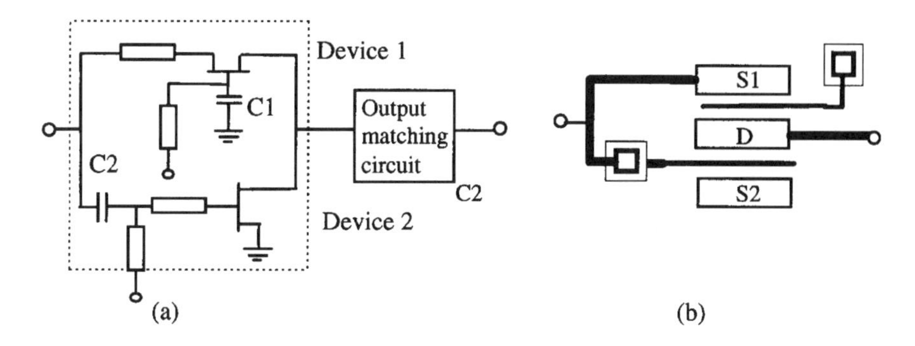

Figure 7.33 Miniaturized frequency doubler, (a) equivalent circuit, and (b) simplified chip layout.

signal at the drain. The phase of the common source device, as previously discussed, is 180 degrees at low frequencies. The harmonics are generated by biasing the devices near the pinch-off voltage. Therefore, combining both drains, all odd harmonics are rejected and even harmonics are added.

The dc isolation between both devices is provided by capacitor C2. The drain bias and dc return for the source of device 1 are done off chip. The realization of this function in MMIC is straightforward as observed in Figure 7.33(b), where the drain is common to both sources.

The transmission lines in series with the gate can be used to compensate for phase differences at the devices output. A frequency doubler built after this approach provided a conversion loss of 10 ± 1 dB over the 3 to 10 GHz fundamental frequency band for an input power of +10 dBm. The LO isolation was better than 15 dB over the same band, which precludes its direct application over the band. It can, however, be useful if the band is divided into sub-bands that can be covered by external filters to remove all unwanted harmonics.

Investigation of higher order MMIC multipliers is now under way. An interesting example of a frequency quadrupler built using HEMT technology, with a topology very similar to the tuned quadrupler of Section 7.3 was recently reported [15]. A 15 to 60 GHz quadrupler provided a multiplication gain of –5 dB at an input power of 0 dBm. Due to the nature of the topology, the bandwidth is narrow, less than 5% for a 3 dB power variation.

7.5.2 Commercial MMICs

One will find on the market amplifier blocks designed for gain, low noise, and medium power, some of which can also be used as multipliers. Although they might not give the best efficiencies, it is a simple way of using MMIC technology for frequency multipliers.

Assume a given MMIC presents the topology illustrated in Figure 7.34, which is ideal from the point of view of general purpose application. There are two ways to employ this MMIC as a multiplier: first, by biasing the first stage as a buffer amplifier, and biasing the second as a multiplier. If the second stage drain is directly accessible, then the concepts previously described are still applicable. All that is required is to connect a transmission-line phase-shifter between the die and the bandpass filter and adjust for best performance.

In this case, it is possible to make a simple model for the MMIC, with bonding wires and calculate the approximate dimensions of the transmission-line phase-shifter. If there is an output matching circuit connected internally to the drain, then the line's phase will have to be determined experimentally. The presence of an output matching circuit on chip will limit the performance, but in general it should still provide usable performance.

The second approach consists in biasing the first stage as a multiplier and the second as an amplifier. This approach works well only if the output stage has a reasonable gain at the desired output harmonic. In this case there is no means to optimize the drain load for best efficiency, since it is limited to the load presented by the output stage. It can, however, perform adequately if the biasing and coupling from first to second stage rejects the fundamental.

A commercial general purpose MMIC amplifier designed to operate from 20 to 40 GHz, type HMMC-5040 from Hewlett-Packard Company, falls into this category [16], and was one of the first commercial MMICs available that could perform both amplifying and multiplying functions. A photo of this MMIC is shown in Figure 7.35. It contains four amplification stages, the first of which consists of a distributed amplifier with two devices. It has a good

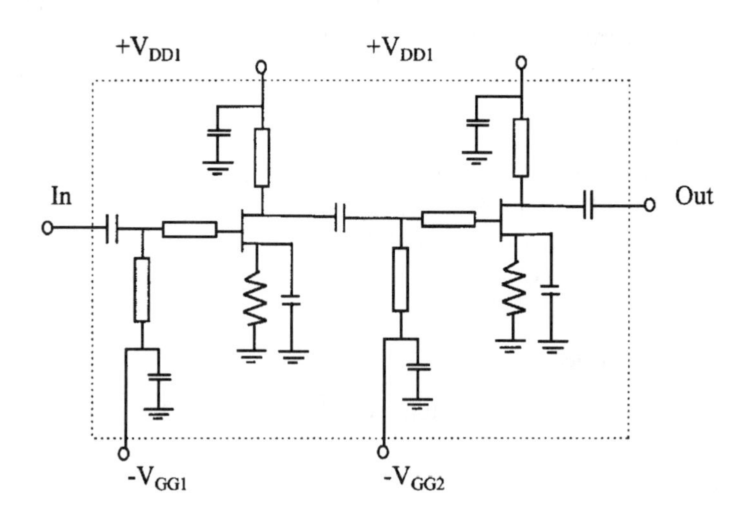

Figure 7.34 Typical two stage MMIC amplifier.

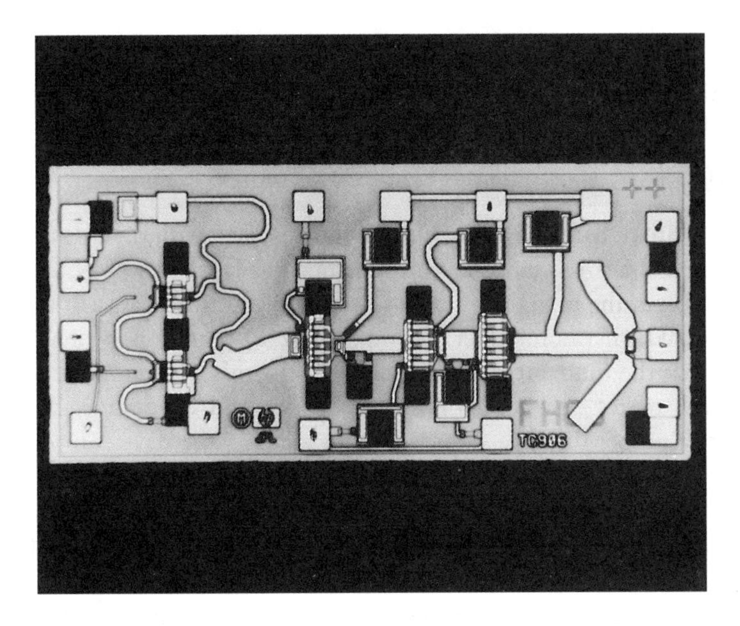

Figure 7.35 Photo of the MMIC-type HMMC-5040 (*Courtesy of Hewlett-Packard Company, Santa Rosa, CA*).

input match from dc to 40 GHz and can operate as an even- or odd-harmonic multiplier by controlling gate and drain bias. Since the coupling between stages cuts off sharply below 20 GHz, the remaining three stages will amplify the harmonics generated by the multiplier that falls in-band. The advantage of this approach is its simplicity, just add a bandpass filter to the MMIC and apply the proper bias. A typical doubler performance when driven by an input power of +14 dBm, is a gain of 6 dB resulting in nearly 20 dBm available power at the output. The trade-off is between the high dc power dissipated, nearly 1.5W, and the cost compared to a discrete approach.

The use of commercial amplifier blocks results in a simple way to build a multiplier. Depending on the die available, the multiplier may perform well from a RF standpoint but may be seriously inefficient from the point of view of dc power consumption. Therefore, it is up to the designer to make the trade-off of using commercial MMICs over custom MMICs or discrete versions.

7.6 Harmonic Oscillators

7.6.1 Single-ended

A topology for an X-band harmonic oscillator has been published [17], making use of a dielectric resonator as the frequency-controlling element. The circuit used a package HEMT, an NE20383A from NEC, as the active element was designed for fundamental oscillation at 9.0 GHz. The circuit schematic using microstrip lines is illustrated in Figure 7.36, and the dimensions for the electrical angles are in Table 7.6.

Drain Circuit

The drain circuit contains a matching circuit cascaded with the harmonic filter. The desired drain reactance at the fundamental frequency is obtained by a combination of the harmonic filter's reactance plus the reactance of the matching circuit. The harmonic filter is composed of 90 degree open-circuited stubs which block the fundamental frequency and the third harmonic. The filter is transparent to the second harmonic frequency and an open-circuited stub placed at a proper distance from the drain matches the device impedance.

Source Circuit

Two open-circuited stubs are connected to the source, one at each lead of a double source package. One of them gives the required capacitive reactance to generate the negative resistance at the fundamental frequency. The other is about 180 degrees longer and stabilizes the device at lower frequencies by presenting a low impedance to ground. At the second harmonic, the bias filter introduces a low impedance to ground.

Gate Circuit

The oscillation condition at the fundamental frequency is obtained by calculating the resonator position from the gate that gives the impedance required by (6.9) and (6.10). At the second harmonic, very low power should be wasted

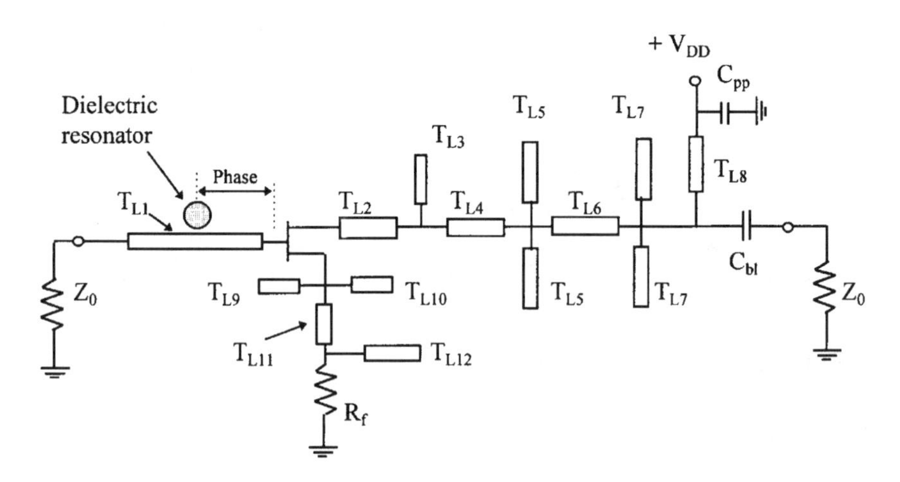

Figure 7.36 Harmonic oscillator schematic diagram.

Table 7.6
Harmonic Oscillator Network Elements
(electric angles normalized at f_0 = 9.0 GHz)

Element No	Element Type	Parameters
T_{L1}	Series line	$Z_0 = 50$ ohm/$\theta = 180°$
Phase	Series line	$Z_0 = 50$ ohm/$\theta = 125°$
T_{L2}	Series line	$Z_0 = 50$ ohm/$\theta = 17°$
T_{L3}	Open stub	$Z_0 = 50$ ohm/$\theta = 28°$
T_{L4}	Series line	$Z_0 = 50$ ohm/$\theta = 17°$
T_{L5}	Open stub	$Z_0 = 50$ ohm/$\theta = 90°$
T_{L6}	Series line	$Z_0 = 50$ ohm/$\theta = 90°$
T_{L7}	Open stub	$Z_0 = 50$ ohm/$\theta = 90°$
T_{L8}	Short stub	$Z_0 = 90$ ohm/$\theta = 45°$
T_{L9}	Open stub	$Z_0 = 30$ ohm/$\theta = 24°$
T_{L10}	Open stub	$Z_0 = 30$ ohm/$\theta = 204°$
T_{L11}	Series line	$Z_0 = 90$ ohm/$\theta = 90°$
T_{L12}	Open stub	$Z_0 = 40$ ohm/$\theta = 90°$
R_f	Resistor	$R = 66$ ohm

at the gate, which is accomplished by placing a 90 degree open-circuited stub at that frequency close to the 50 ohm load. As a result, the second harmonic power will increase without perturbing the fundamental frequency of oscillation and introducing negligible effects on the circuit stabilization at low frequencies.

A circuit like this, using a $0.3 \times 280\ \mu m^2$ gate device, was constructed on 10 mils thick Duroid® substrate from Rogers. It provided more than +6 dBm output power at 18 GHz with a 6% drain efficiency. The oscillator phase noise at 100 kHz from the carrier was equal to 95 dBc/Hz.

7.6.2 Push-push Version

The equivalent "balanced" version for this circuit is the topology known as a "push-push" oscillator, for a long time used by bipolar designers [18]. The FET version is described in Figure 7.37. The push-push oscillator is a frequency doubling oscillator employing two transistors, each oscillating at one-half the desired output frequency. The transistors oscillate out-of-phase with respect to each other, causing the fundamental frequency to cancel and the second harmonic to add in phase. Push-push designs have several advantages over other topologies. Designing at one-half the frequency increases the resonator Q, decreases the parasitics that are encountered, and extends the useful frequency range of transistors. The topology described next uses a dielectric resonator to

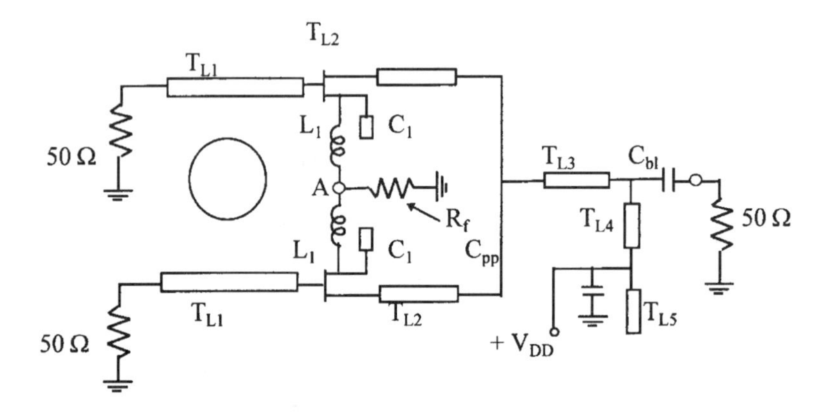

Figure 7.37 FET Push-push oscillator.

perform the function of a resonator and also guarantees the push-push operation by locking the phase of each oscillator at a 180 degree phase difference.

The circuit analysis starts by recognizing that this topology presents two modes of operations, namely differential or *odd-mode*, and common or *even-mode*. The odd mode is guaranteed by a dielectric resonator coupled to the two gate lines [19], whose cross-section is represented in Figure 7.38, showing the current lines, (i.e., the magnetic field). Since the $TE_{01\delta}$ is the dominant mode for the dielectric resonator, two magnetic dipoles exist, which are exactly of opposite phase at the resonant frequency.

In the differential mode, point A of Figure 7.37 is a virtual ground so that the oscillator circuit can be simplified by the device and its feedback capacitance, C_1, as shown in Figure 7.39. The wire bond inductance, L_1, was suppressed from the circuit since it only has a dc return function in this mode. Proper device biasing has to trade-off a high gain to satisfy the conditions for start-up of oscillations, and a high generation of even harmonics in the drain current, usually requiring class AB operation.

The start up of oscillations has to be valid in this mode of operation. Therefore, impedance Z_A must present a negative real part greater than the

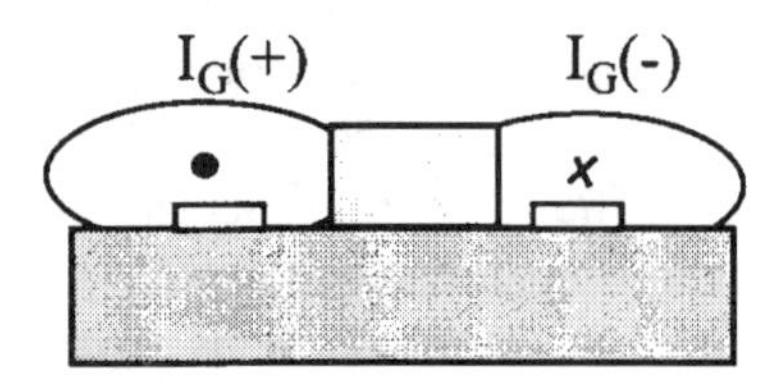

Figure 7.38 H-fields created by the Dielectric resonator centered between two parallel transmission lines.

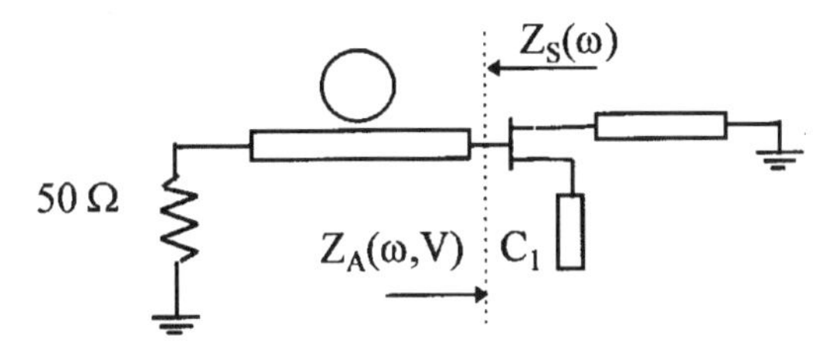

Figure 7.39 Odd-mode equivalent circuit.

losses introduced by the dielectric resonator. The reactances from each circuit must cancel at the oscillation frequency.

$$|-R_A| > R_S$$
$$X_A + X_S = 0$$

In the common-mode of operation, there should be no possibility of oscillations. In this mode, point A is not longer grounded, so that a symmetry can be created if common impedances are doubled, and the circuit is halved in the manner shown in Figure 7.40. Therefore, an additional condition has to be added to the previous one.The source resistor in this mode must neutralize the negative resistance generated by the capacitive feedback element.

A topology similar to the one described [19], provided +3.0 dBm output power around 33 to 36 GHz. The resulting phase noise at 100 kHz from the carrier was nearly 100 dBc/Hz. The performance of such an oscillator is a function of the electrical symmetry provided by the physical circuitry. Therefore, identical devices as well as careful circuit layout are important in this topology. These requirements make this circuit an ideal option for MMIC application; however, the gates have to be coupled to a dielectric resonator off-chip. A complete integrated version can be obtained by eliminating the dielectric resonator and connecting both gates with a half wavelength line at the fundamental frequency, thus guaranteeing a phase shift of 180 degrees between each gate.

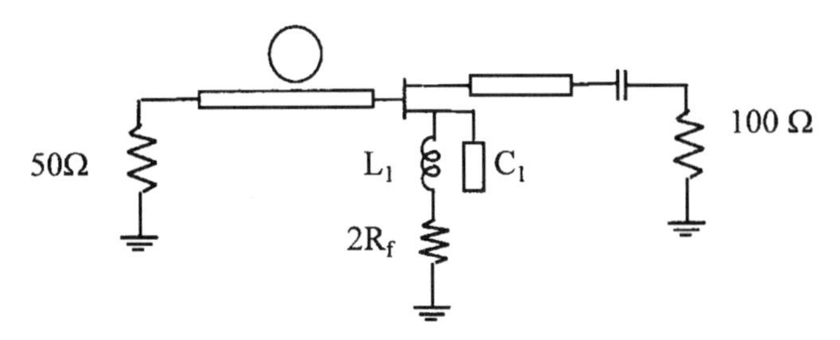

Figure 7.40 Even-mode equivalent circuit.

References

[1] Pan, J. J., "Wideband MESFET Microwave Frequency Multiplier," *IEEE MTT-S 1978 International Microwave Symposium Digest,* 1978, pp. 306–308.

[2] Camargo, E., R. Perichon, R. Soares, and M. Goloubkoff, "Sources of Non-linearity in GaAs MESFET Frequency Multipliers," *IEEE MTT-S 1983 International Microwave Symposium Digest,* Boston, MA, 1983, pp. 343–345.

[3] Mantovani, F, "Active MESFET Multipliers Solve Low Signal Levels," *Microwaves & RF,* Vol No. 8, August 1984, pp. 129–131.

[4] Iyama, Y., A. Iida, T. Takagi, and S. Urasaki, "Second-Harmonic Reflector Type High Gain FET Frequency Doubler Operating in K-Band," *IEEE MTT-S 1989 International Microwave Symposium Digest,* Long Beach, CA, 1989, pp. 1291–1294.

[5] Rauscher, C., "High-Frequency Doubler Operation of GaAs Field-Effect Transistors," *IEEE Transactions on Microwave Theory and Techniques,* Vol 31, No. 6, June 1983, pp. 462–473.

[6] Henkus, J. C., R. Overduin, and P. J. Koomen, "A Wideband Tripler for X-Band in Microstrip," *Microwave Journal,* Vol No. 3, March 1993, pp. 106–111.

[7] Gilmore, R., "Concepts in the Design of Frequency Multipliers," *Microwave Journal,* Vol No. 3, March 1987, pp. 129–139.

[8] Stancliff, R., "Balanced Dual Gate GaAs FET Frequency Doublers," IEEE MTT-S 1981 International Microwave Symposium Digest, Long Beach, CA, 1981, pp. 143–145.

[9] Henderson, B. C. and A. W. Denning, "Single-Balanced Mixer Forms Dual MM-Wave Frequency Multiplier," *Microwaves & RF,* Vol No. 10, October 1990, pp. 81–86.

[10] Lott, U., "Low Loss MESFET Frequency Quadrupler From 5 to 20 GHz." Proceedings of the 21st European Microwave Conference, Stuttgart, Germany, September 1991, pp. 1502–1506.

[11] Stein, O. Von and J. Sherman, "Odd Order MESFET Multipliers with Broadband, Efficient, Low Spurious Response," *IEEE MTT-S 1996 International Microwave Symposium Digest,* 1996, San Francisco, CA, pp. 667–670.

[12] Funabashi, M., T. Inoue, K. Maruhashi, K. Hosoya, M. Kuzuhara, K. Kanekawa, and Y. Kobayashi, "A 60 GHz MMIC Stabilized Frequency Source Composed of a 30 GHz DRO and a Doubler," *IEEE MTT-S 1995 International Microwave Symposium Digest,* Orlando, FL, 1995, pp. 71–74.

[13] Hirota, T. and H. Ogawa, "Uniplanar Monolithic Frequency Doublers," *Transactions on Microwave Theory and Techniques,* Vol 37, No. 8, August 1989, pp. 1249–1254.

[14] Hiraoka, T., T. Tokumitsu, and M. Akaike, "A Miniaturized Broad-Band MMIC Frequency Doubler," *Transactions on Microwave Theory and Techniques,* Vol 38, No. 12, December 1990, pp. 1932–1937.

[15] Shirakawa, K., Y. Kawasaki, Y. Ohashi, and N. Okubo, "A 15/60 GHz One-Stage MMIC Frequency Quadrupler," Microwave And Millimeter-Wave Monolithic Circuits Symposium, 1996, pp. 35–38.

[16] "HMMC-5040 As a 20–40 GHz Multiplier," Application Note # 50 by Hewlett-Packard.

[17] Tupynamba, R., E. Camargo, and F. S. Correra, "A HEMT Harmonic Oscillator Stabilized by an X-Band Dielectric Resonator," *MTT-S 1991 International Microwave Symposium Digest,* Boston, June 1991, pp. 277–280.

[18] Bender, R., and C. Wong, "Push-Push Design Extends Bipolar Frequency Range," *Microwaves & RF,* Vol No. 10, October 1983, pp. 91–98.

[19] Pavio, A. and M. A. Smith, "Push-Push Dielectric Resonator," *MTT-S 1985 International Microwave Symposium Digest,* Saint Louis, MO, June 1985 pp. 266–269.

Appendix A:

Reference Device

The device used in all simulations carried out in this book is the AT-10600 manufactured by Avantek, now Hewlett-Packard Company. This GaAs FET chip has a nominal 0.3 μm-gate length with a total gate periphery of 250 μm. The device data sheet recommends its application in low noise applications due to its low noise figure, 1.8 dB @ 12 GHz with 9.0 dB associated gain and in power applications where +18 dBm @ 18 GHz can be obtained. Other important parameters obtained from the Data book, are in Tables A.1 and A.2. Noise figure parameters are on Table A.3 and S-parameters in Tables A.4 and A.5 for two drain-current biases.

Table A.1
Absolute Maximum Ratings

Parameter	Symbol	Absolute Maximum
Drain-Source Voltage	V_{DS}	+7.0 V
Gate-Source Voltage	V_{GS}	–4.0 V
Drain Current	I_{DS}	I_{DSS}
Power Dissipation	P_T	275 mW
Channel Temperature	T_{CH}	+175 °C
Thermal Resistance	θ_{jc}	225 °C/W

Table A.2
Electrical Specifications, T_A = 25 °C

Symbol	Parameters and Test Conditions	Units	Typical
NFmin	Optimum Noise Figure—f = 8.0 GHz	dB	1.5
	@V_{DS} = 3V/I_{DS} = 10 mA—f = 12.0 GHz	dB	1.8
	f = 14.0 GHz	dB	2.0
GA	Gain @ NFmin—f = 8.0 GHz	dB	12.0
	f = 12.0 GHz	dB	9.0
	f = 14.0 GHz	dB	8.0
P 1dB	Output Power—f = 12.0 GHz	dBm	8.0
	V_{DS} = 5.0 V/I_{DS} = 30 mA—f = 12.0 Ghz	dB	8.0
	1 dB compressed gain		
gm	Transconductance V_{DS} = 3V/V_{GS} = 0V	mS	40.0
V_P	Pinch-off Voltage V_{DS} = 3V/I_{DS} = 1mA	V	–1.5
I_{DSS}	Saturated Drain Current V_{DS} = 3V/V_{GS} = 0	mA	50.0

Table A.3
Noise Parameters, V_{DS} = 3V/I_{DS} = 10 mA

Frequency GHz	NFmin	Gamma Opt Mag	Gamma Opt Ang	Rn/50
6.0	1.3	.67	52	1.08
8.0	1.5	.45	84	1.03
12.0	1.8	.22	–158	.84
14.0	2.0	.33	–136	.91

Table A.4

Typical S-Parameters: Common Source, Z_0 = 50 Ω T_A = 25 °C @ V_{DS} = 3V, I_{DS} = 10 mA

	S_{11}		S_{21}		S_{12}		S_{22}	
Freq GHz	**Mag**	**Angle**	**Mag**	**Angle**	**Mag**	**Angle**	**Mag**	**Angle**
2.0	0.98	–19	2.11	161	0.034	82	0.68	–4
3.0	0.96	–28	2.11	151	0.051	78	0.67	–6
4.0	0.93	–37	2.15	142	0.084	69	0.64	–9
5.0	0.90	–47	2.22	131	0.067	74	0.61	–13
6.0	0.86	–59	2.27	120	0.101	64	0.57	–17
7.0	0.80	–72	2.29	110	0.117	59	0.52	–23
8.0	0.74	–83	2.32	100	0.130	54	0.48	–28
9.0	0.68	–97	2.33	89	0.143	48	0.43	–36
10.0	0.61	–114	2.31	77	0.156	42	0.38	–46
11.0	0.55	–133	2.27	66	0.166	35	0.33	–56
12.0	0.52	–153	2.18	54	0.174	28	0.28	–66
13.0	0.49	–74	2.10	44	0.180	21	0.24	–78
14.0	0.49	165	2.00	33	0.184	10	0.17	–99
15.0	0.52	144	1.84	21	0.186	6	0.15	–109
16.0	0.54	131	1.69	12	0.187	3	0.13	–121
17.0	0.57	117	58	3	0.189	–6	0.10	–177
18.0	0.58	105	1.45	–6	0.190	–12	0.13	145

Table A.5

Typical S-Parameters: Common Source, Z_0 = 50 Ω T_A = 25 °C @ V_{DS} = 5V, I_{DS} = 30 mA

	S_{11}		S_{21}		S_{12}		S_{22}	
Freq GHz	**Mag**	**Angle**	**Mag**	**Angle**	**Mag**	**Angle**	**Mag**	**Angle**
2.0	0.98	–22	3.01	158	0.022	85	0.71	–3
3.0	0.93	–34	2.98	146	0.029	81	0.70	–5
4.0	0.87	–45	3.02	136	0.039	78	0.68	–8
5.0	0.81	–59	3.05	124	0.051	76	0.65	–11
6.0	0.74	–72	2.98	112	0.058	74	0.62	–14
7.0	0.68	–86	2.86	101	0.068	72	0.60	–18
8.0	0.63	–100	2.80	91	0.075	70	0.57	–22
9.0	0.56	–113	2.76	82	0.082	65	0.53	–26
10.0	0.51	–130	2.62	72	0.096	68	0.51	–30
11.0	0.46	–143	2.52	62	0.100	66	0.47	–40
12.0	0.42	–156	2.45	54	0.105	64	0.46	–49
13.0	0.41	–168	2.37	46	0.114	60	0.44	–54
14.0	0.40	164	2.29	38	0.129	56	0.42	–60
15.0	0.41	145	2.15	29	0.143	52	0.40	–65
16.0	0.44	132	2.04	21	0.167	45	0.13	–72
17.0	0.47	120	1.97	11	0.185	40	0.10	–80
18.0	0.51	104	1.83	2	0.187	34	0.13	–93

The measured $I_{DS} - V_{DS}$ plots for a typical device is represented in the next figure, for V_{GS} ranging from −1.5 to +0.5 volts. The obtained I_{DSS} is equal to 62 mA @ V_{DS} = 3.0 volts.

The measured $I_{DS} - V_{GS}$ plot for the same device is shown in Figure A.2, for V_{DS} = 3.0 V. Notice that there exists a smooth square law region between V_{GS} = −0.75 volts and −1.5 volts, and a nearly linear relation from V_{GS} = −0.75 to + 0.5 volts.

The measured $I_{GS} - V_{GS}$ plot is depicted in the next figure. It shows a leakage gate current of 1.5 μA for gate voltages up to 0.5 volts and −1.5 volts. The gate voltage can be more negative, up to −4.0 volts, where avalanche current of the gate-to-source diode starts to flow. However, it is not advisable to apply more than 0.5V at the gate if reliability is to be maintained.

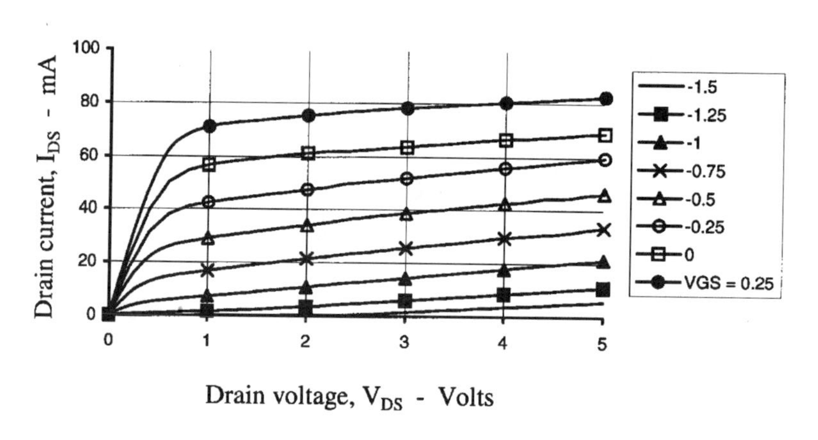

Figure A.1 I_{DS} as a function of V_{DS} and V_{GS}.

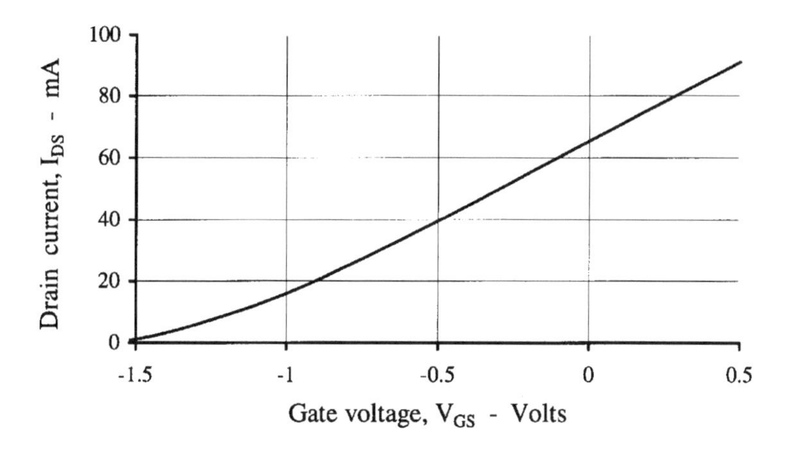

Figure A.2 I_{DS} as a function of V_{GS} @ V_{DS} = 3.0 volts.

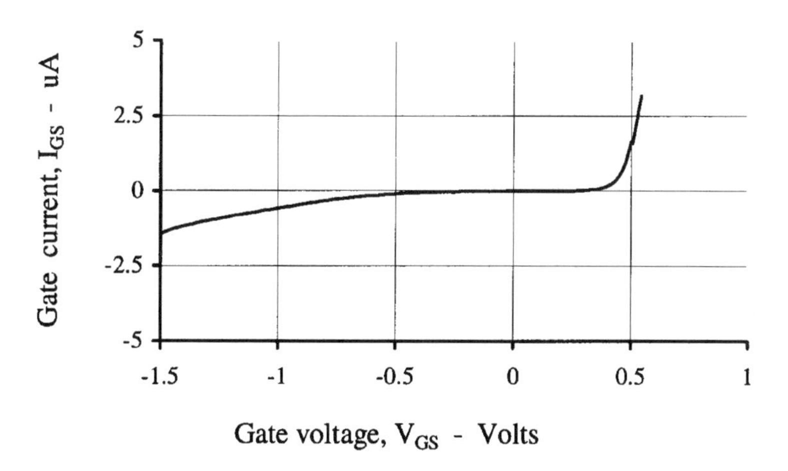

Figure A.3 Gate current as a function of gate voltage, @ V_{DS} = 2.0V.

Appendix B:

Matching Networks

The primary consideration in the design of matching networks is the definition of the technology employed to realize the circuit elements. At low frequencies, up to a few GHz, discrete chip R, L, and C components in surface mount technology (SMT) are the most popular in the industry. In the microwave and millimeter wave range, up to 40 GHz, microstrip lines are the preferred means to build passive circuits in MIC technology. At higher frequencies, the traditional waveguides are used to build components. In MMIC technology, cost is proportional to chip area, therefore lumped elements are used as much as possible for the minimum area they require. The advantage of microstrip elements in frequency multipliers is on their impedance-frequency relation, for instance, a quarter-wave line at the fundamental frequency becomes a half-wave line at the second harmonic, which makes them useful to separate harmonic frequencies. Some examples of matching networks employing microstrip lines are provided in Chapter 7. In this appendix, emphasis is on basic lumped elements topologies to be used in MMIC technology.

B.1 Single L Network [1]

In this class of matching network, depicted in Figure B.1, a common methodology is to place the reactance in parallel with the highest resistor value. The parallel reactance, X_p, transforms the generator impedance R_g, to R_{in} in series with a reactance which is canceled by the series reactance, X_s. The matching condition is that the Q-factor of the parallel circuit equals the Q-factor of the series circuit.

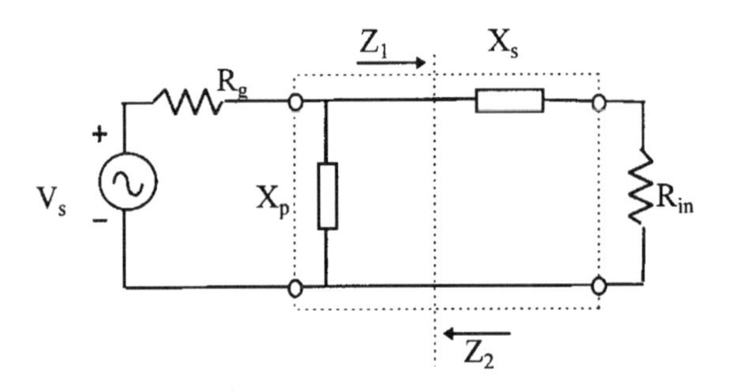

Figure B.1 Matching of R_{in} to R_g.

The matching elements are determined by (B.1) through (B.2) and (B.3). Note that the transformation ratio, $n_t = R_g/R_{in}$, determines the Q-factor. Also note that X_s and X_p may be positive or negative, but they are always opposed to each other.

$$X_p = \pm R_g / Q_p \tag{B.1}$$

$$X_s = \mp R_{in} Q_s \tag{B.2}$$

$$Q = Q_s = Q_p = \sqrt{n_t - 1} \tag{B.3}$$

A better visualization of the matching properties of this network can be obtained from its plot on the Z-plane. For that purpose, let us split the network in two; one containing the series reactance and the other the parallel one. At this point, let us also assume that the series element is an inductance and the parallel one a capacitance. The impedance as a function of frequency for each network is illustrated in Figure B.2.

Observe in this plot that there is only one solution for a perfect matching. It is also observed that mismatching increases rapidly around ω_o. As a matter of fact, the bandwidth is dependent on transformation ratio, n_t, and the device

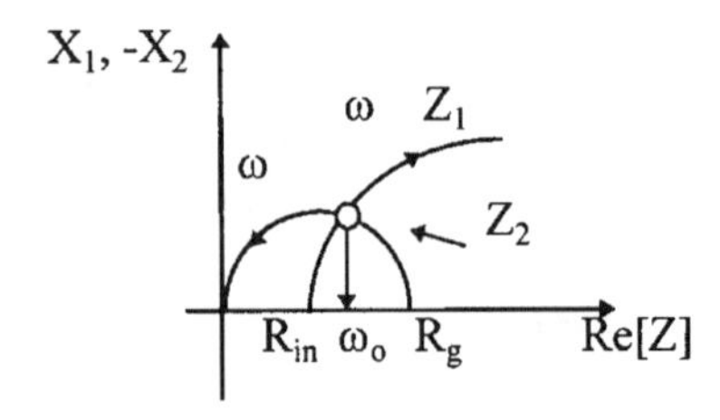

Figure B.2 Matching condition in the Z-plane for the L network.

quality factor, Q. The higher those parameters, the lower the bandwidth, as depicted from Figure B.3, which shows the VSWR as a function of n_t for a network transforming resistive to resistive impedance.

B.2 Double L Network

It is clear from Figure B.3 that bandwidth is greatly reduced if the transformation ratio is greater than 2. Another procedure to improve bandwidth is to use a double L network as shown in Figure B.4, where an intermediate impedance is chosen between both terminations, $R_{\text{in}} < R_{int} < R_g$, reducing the transformation ratio of each, therefore increasing the bandwidth.

The intermediate resistor can be determined as the geometric mean of both impedances, $R_{int} = \sqrt{R_g R_{\text{in}}}$. The Q-factor for both sections are equal, $Q_1 = Q_2$, and given by (B.4) and (B.5), which were derived from a parallel-series transformation.

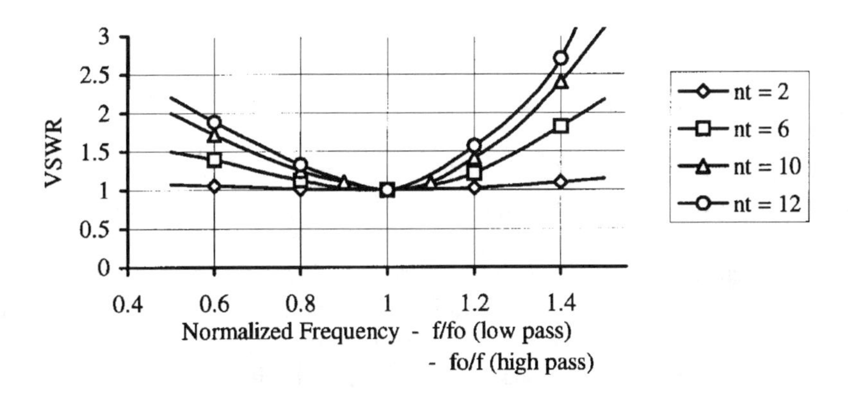

Figure B.3 Bandwidth as a function of n_t [2].

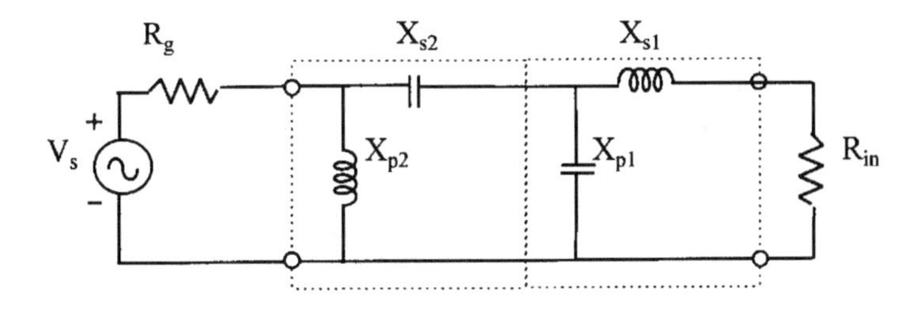

Figure B.4 Double L matching network.

$$Q_1 = \sqrt{R_{int}/R_{\text{in}} - 1} \tag{B.4}$$

$$Q_2 = \sqrt{R_g/\text{R}_{int} - 1} \tag{B.5}$$

The reactance of each circuit element is given by the quality factor of each R-C and R-L network. The circuit is designed to transform R_g to R_{in}, and the series inductance X_{s1} can absorb the inductance required to resonate the gate input capacitance.

$$X_{s1} = \omega L_s = R_{\text{in}}\, Q_1 \tag{B.6}$$

$$X_{p1} = \frac{1}{\omega C_p} = R_{int}/Q_1 \tag{B.7}$$

$$X_{s2} = \frac{1}{\omega C_s} = R_{int}\, Q_2 \tag{B.8}$$

$$X_{p2} = \omega L_p = R_g/Q_2 \tag{B.9}$$

B.3 "PI" Network

A popular network used for impedance matching is the "PI" circuit in its lowpass form, as shown in Figure B.5 (i.e. two parallel capacitors and one series inductor). The design equations are obtained by calculating the impedance at the generator port and equating the imaginary part to zero and the real part to R_g. Two equations result from this procedure and a third one is required to determine all three elements. The third condition is the Q-factor for the $R_g C_1$ parallel circuit which is made as small as possible for maximum bandwidth and minimum mismatch.

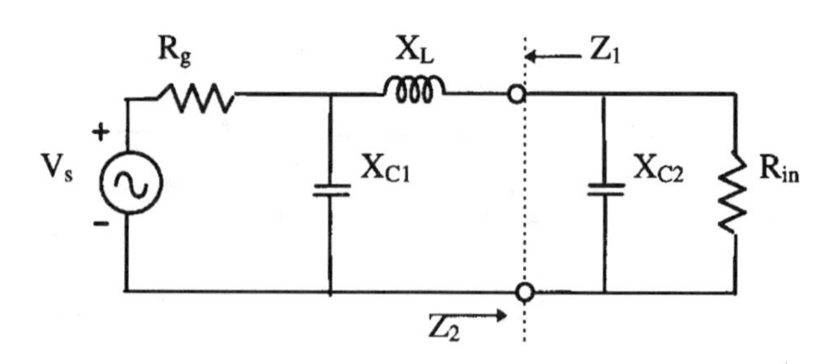

Figure B.5 The "PI" topology.

The required equations in function of the circuit parameters are described below and the first step in this approach consist in the definition of a Q-factor.

$$X_{C1} = R_g / Q \tag{B.10}$$

$$X_{C2} = \sqrt{\frac{R_{\text{in}}}{(R_{\text{in}}/R_g)(Q^2 + 1) - 1}} \tag{B.11}$$

$$X_L = \frac{QR_g}{Q^2 + 1}\left[1 + \frac{R_{\text{in}}}{QX_{C2}}\right] \tag{B.12}$$

The plot of the impedances, Z_1 and Z_2 on the Z-plane, is depicted in Figure B.6. The impedance Z_1 corresponds to a parallel RC circuit, whose plot is a half circle as a function of frequency. The impedance Z_2 presents a more complex variation with frequency.

The matching condition is obtained when both plots are tangent to each other, at the center frequency of operation, ω_o. Comparing this matching condition with the one from Figure B.2, one can observe that the bandwidth for this topology is larger.

B.3 Modified "L" Network

This topology, shown in Figure B.7, applies reactance compensation to improve bandwidth. An example is demonstrated for the match of gate impedance. The circuit is divided into two parts: one where the input gate capacitance is resonated so that the input impedance is resistive at the center frequency of operation, and the second part is an "L" network employed to transform the resulting resistance to the generator impedance. A parallel LC circuit resonant at the center frequency of operation is added to compensate the reactance variation of the resonated gate impedance.

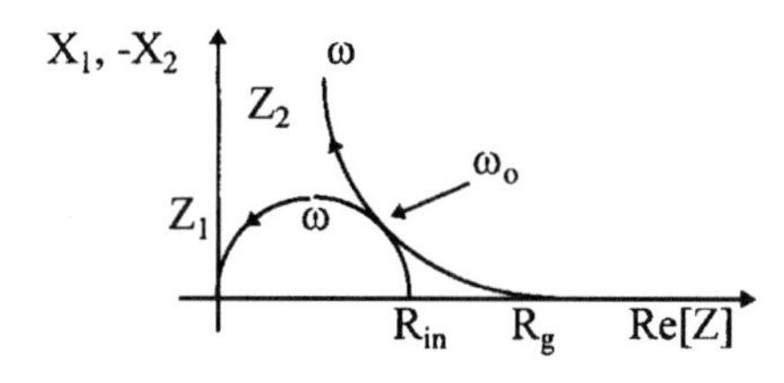

Figure B.6 Matching condition "PI" network.

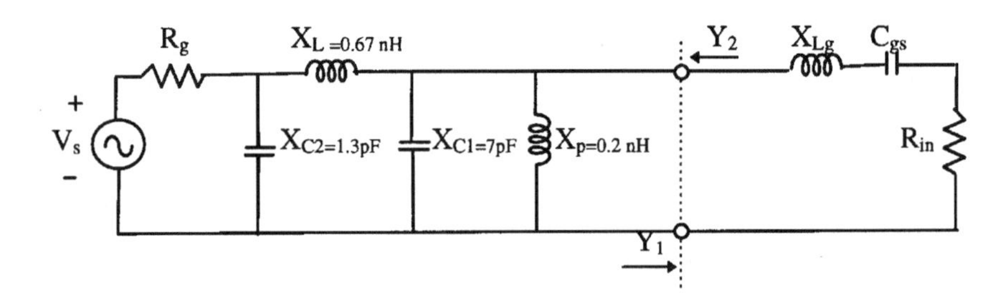

Figure B.7 The modified "L" topology.

To determine the element values, the matching conditions [3,4], (B.13) to (B.14), and compensation conditions (B.15), must be verified. The derivation of these conditions analytically is complex, so that computer optimization is the adequate tool to determine the elements.

$$Re[Y_1] = Re[Y_2] \tag{B.13}$$

$$Im[Y_1] = -Im[Y_2] \tag{B.14}$$

$$Q_1 = -Q_2 \tag{B.15}$$

where,

$$Q = \left| \frac{\overline{\omega}_0}{2G} \frac{dB}{d\omega} \right| \text{ at } \omega = \omega_0$$

The simulation results for a typical MESFET gate impedance, R_{in} = 10 ohm and C_{gs} = 1 pF, matched for the center frequency of 5 GHz, are shown in the plots of Figures B.8 and B.9. The first plot depicts the

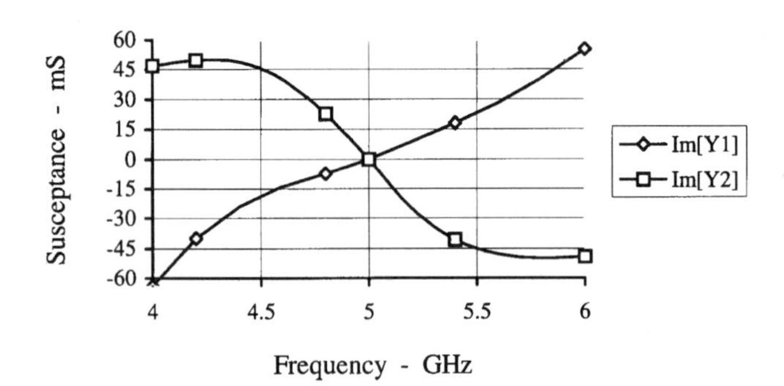

Figure B.8 Reactance versus frequency.

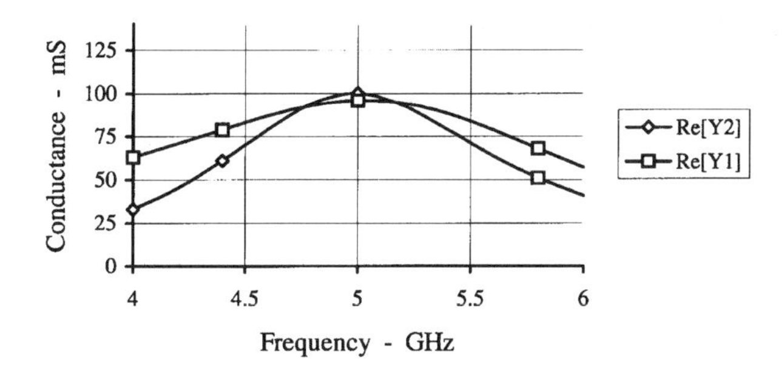

Figure B.9 Resistance versus frequency.

reactance compensation of each half showing the opposite reactance sign of each as a function of frequency.

Observe the compensation of reactances within the range 4 to 6 GHz, corresponding to more than 40% fractional bandwidth. The second plot illustrates the resistive matching over the same band.

The plot shows a nearly perfect match from 4.5 to 5.5 GHz, and a mismatch at the bandwidth extremes. Therefore, if the targeted VSWR is less than 1:1.5, then the bandwidth is reduced to 20% which still is an adequate value for a large number of microwave circuits.

References

[1] Besser, L., "Reactive Transformation of Resistances," *Applied Microwaves,* Winter 1993, pp. 104–110.

[2] Soares, R, *GaAs MESFET Circuit Design*, Norwood, MA: Artech House, 1988, pp. 320.

[3] Aitchinson, C. S. and R. V. Gelsthorpe, "A Circuit Technique for Broadbanding the Electronic Tuning Range of Gunn Oscillators," *IEEE Journal of Solid-State Circuits,* Vol. SC-12, No. 1, Feb. 1977.

[4] Camargo, E., D. Consoni, and R. Soares, "Reactance Compensation Matches FET Circuits," *Microwave & RF,* June 1985, pp. 93–95.

Appendix C:

Transferring Internal (I,V) to External Terminals

The transfer of currents and voltages of an intrinsic MESFET model is carried out step by step, starting with the simplified device model. The linear parasitic elements are described by either "TEE" or "PI" networks, sequentially connected to the device model. For instance, three parallel conductances, Y_1, Y_2 and Y_3, in the form of a "PI" network are added to the three terminal device of Figure C.1.

The internal set of voltages and currents (I,V) are transferred to the external terminals (I′,V′), by means of the following matrix equation:

$$\begin{bmatrix} V_1' \\ V_2' \\ I_1' \\ I_2' \end{bmatrix} = \begin{bmatrix} 1 & 0 & 0 & 0 \\ 0 & 1 & 0 & 0 \\ Y_1 + Y_3 & -Y_3 & 1 & 0 \\ -Y_3 & Y_2 + Y_3 & 0 & 1 \end{bmatrix} \cdot \begin{bmatrix} V_1 \\ V_2 \\ I_1 \\ I_2 \end{bmatrix} \tag{C.1}$$

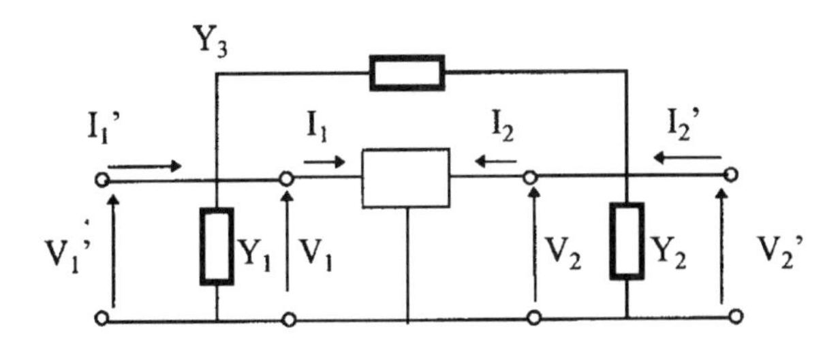

Figure C.1 Transfer of I,V for "PI" network.

In the case of a "TEE" network, three series impedances, Z_1, Z_2 and Z_3, are added to the three terminal device of Figure C.2.

The internal set of voltages and currents (I,V) in the "TEE" network are transferred to the external terminals (I′,V′), by means of the following matrix equation:

$$\begin{bmatrix} V_1' \\ V_2' \\ I_1' \\ I_2' \end{bmatrix} = \begin{bmatrix} 1 & 0 & Z_1 + Z_3 & Z_3 \\ 0 & 1 & Z_2 & Z_2 + Z_3 \\ 0 & 0 & 1 & 0 \\ 0_3 & 0 & 0 & 1 \end{bmatrix} \cdot \begin{bmatrix} V_1 \\ V_2 \\ I_1 \\ I_2 \end{bmatrix} \tag{C.2}$$

This is a simple method of transferring voltages and currents, and to take account of the parasitics associated with a device. It is a linear procedure which can be calculated by means of simple software program.

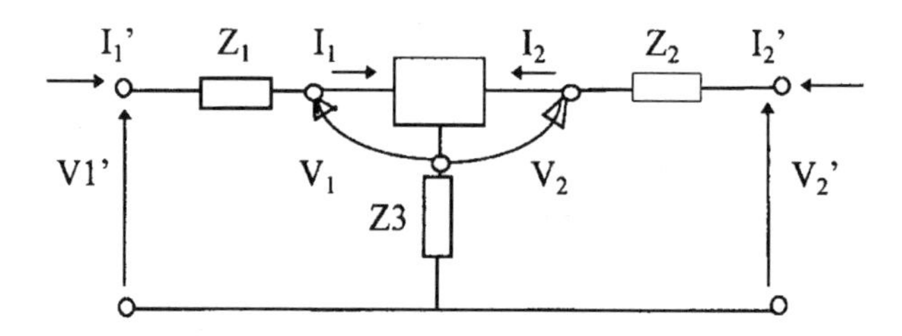

Figure C.2 Transfer of I,V for "TEE" network.

Appendix D:

Characterization of Frequency Multipliers

The set-up of Figure D.1 is adequate for a complete characterization of frequency multipliers. It contains a sweep generator connected to a scalar network analyzer (SNA), which sweeps reflected signals within the fundamental frequency bandwidth. The signal from the generator is sampled by two directional couplers. The first is used as a power reference and the second to measure the circuit return loss. At the output of the multiplier there is another coupler connected to a spectrum analyzer (SA), which has the purpose of detecting any instability present in the circuit, as well as to measure the phase noise

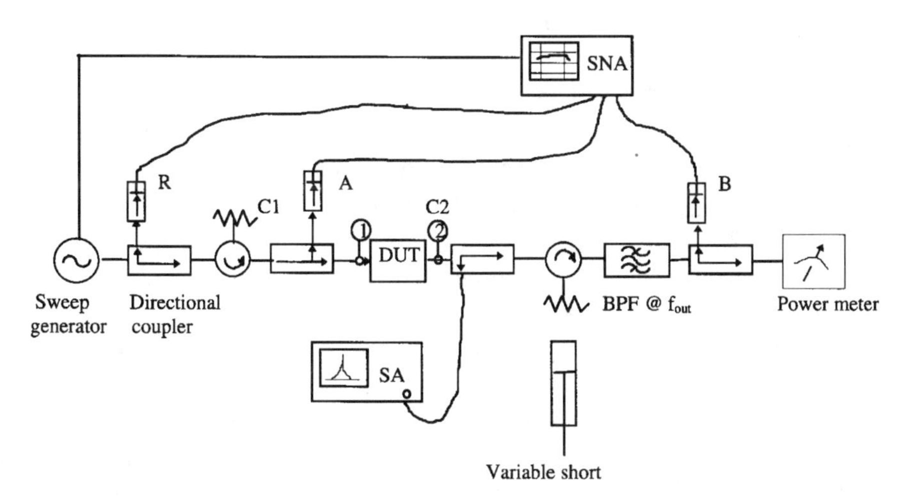

Figure D.1 Multiplier measurement set-up.

degradation. The signal is then fed to an isolator and a bandpass filter tuned at the multiplier output. In this way, the power meter and output detector will be measuring only the desired signal, without influence of other harmonic signals in the measurement.

Before starting any measurements, it is necessary to calibrate the test set-up to take account of losses in the measurement. The SNA should be set to dc mode, avoiding application of a modulated signal into a non-linear circuit.

1. The first calibration can be done by connecting the power meter in place of the Device Under Test (DUT) at port 1. That will allow calibration of input power on the DUT, by calibrating the reading of diode R on the SNA. It is important that the power at that point is leveled within 0.5 dB within the frequency band, to minimize errors in the multiplication gain measurement. If necessary, power leveling can be obtained by adding a coupler and crystal detector before the input coupler and connecting the detected dc level to the ALC loop provided by the sweep generator.
2. The return loss measurement is calibrated by connecting a short and open at port 1 and normalizing the power reference in the channel A of the SNA.
3. The power meter is connected back at the output. The generator sweeping now at the output frequency band has its output power adjusted to the same level read at port I. The generator is then connected to port II, and the SNA is normalized to compensate losses over frequency of the output hardware in channel B. The total losses are read from the power meter.
4. The SA can also be calibrated at this point by offsetting the power read in the power meter with the power indicated by the SA.
5. The generator is connected back at the input coupler and the DUT is connected to the test ports. The SNA is then switched to the B/R mode to read output power referenced to input power.

The generator should be returned to the previous position and the DUT, connected as illustrated in the figure. The power supplies are not shown in this diagram. The system is now ready to be used in the characterization of frequency multipliers. The first set of measurements to be taken from a multiplier is the multiplication gain and input return loss over the frequency band. The power performance is next, and is carried out by setting the generator to CW, and to zero sweep, (i.e. $\Delta f = 0$. The generator is then switched to the power sweep mode). The plot on the SNA will display the gain performance as a function of input power. The SA will display the output signal and also the level of unwanted harmonics. If the SA model is adequate, phase noise after the multiplication can be read directly from its screen.

To detect the circuit stability, a variable short is connected in place of the bandpass filter. By changing the short position one can observe on the SA the circuit stability. This test should also be carried out at the input, by connecting the variable short at port 1 and observing the eventual presence of oscillations.

List of Symbols

A_0, A_1, A_2, A_3	Drain current coefficients, Curtice Cubic model
A_{cn-1}, A_{cn+1}	Power coupling or reflection coefficient at the harmonics $(n-1)$, $(n+1)$
α	Adjust knee voltage in Statz-Pucel model
α_M	Adjust knee voltage in Materka model
BJT	Bipolar junction transistor
B	Adjust $I_{DS} - V_{DS}$ law in Statz-Pucel
BV_{GD}	Gate-to-drain breakdown voltage
BV_{GS}	Gate-to-source breakdown voltage
BV_{DS}	Drain-to-source breakdown voltage
β	DC parameter to adjust DC gain
β_{in}	Coefficient representing input phase modulation
β_{out}	Coefficient representing output phase modulation
β_m	Frequency modulation index
β_c	Adjust pinch-off voltage in Curtice Cubic model
C_{bl}	DC blocking capacitor
C_{bp}	RF bypass capacitor
C_{gs}	Gate-to-source capacitance
C_{gd}	Gate-to-drain capacitance
C_{ds}	Drain-to-source capacitance
C_{GS0}	Gate capacitance when $V_{GS} = 0$
C_{eq}	Equivalent gate capacitance

C_o	Center of stability circle
CG	Conversion gain
DRO	Dielectric Resonator Oscillator
D_{GS}	Gate-to-source diode
D_{GD}	Gate-to-drain diode
δ	Smootl•ing parameter for capacitance in Statz-Pucel model
δ_T	Gate modulation parameter
$\Delta\omega$	Angular frequency band
Δf_{peak}	Maximum frequency deviation
η	Drain efficiency $\eta = P_{\text{out}}/P_{DC}$
f_c	Signal carrier frequency
f_m	Modulating frequency or offset frequency from carrier
f_T	Transition frequency; frequency where current gain is unity
$f_{\max}$	Maximum oscillation frequency
$\phi°$	Electrical angle of a transmission line in degrees
$\phi(t)_p$	Instantaneous angle modulation
$\phi(S_{11})$, $\phi(S_{22})$	Phase of parameters S_{11} and S_{22}, in radians
γ_T	Adjust V_P modulation in TOM model
Γ	Voltage reflection coefficient
g_m	Small signal transconductance
G_M	Large signal transconductance
G_{M0}	Average large signal transconductance
G_{Mn}	Multiplication transconductance of nth order
g_{ds}	Small signal output conductance
G_{DS}	Large signal output conductance = $1/R_{DS}$
GaAs	Gallium Arsenide
HBT	Heterojunction bipolar transistor
HEMT	High Electron Mobility Transistor
I_{11}	Fundamental frequency current at terminal 1
I_{21}	Fundamental frequency current at terminal 2
I_{12}	Second harmonic frequency current at terminal 1
I_{22}	Second harmonic frequency current at terminal 2
IL	Insertion loss
I_s	Reverse saturation current of the Schottky barrier
$I_n(x)$	Bessel function of the first kind, or order n and argument x

I_{DS}	DC drain-source current
I_{DSS}	Drain current at $V_{GS} = 0$
I_G	DC gate current
I_{GS}	DC gate-to-source current
I_{GD}	DC gate-to-drain current
I_{VB}	Leakage current due to breakdown voltage of the Schottky barrier
I_F	Maximum drain current
$I_{ds}(t)$	Total drain current, $I_{ds}(t) = I_{D0} + I_{dn}\cos\omega t$
I_{D0}	Drain bias current
I_{dn}	Drain current at the nth harmonic
I_p	Peak amplitude of $I_{ds}(t)$
k	Boltzmann constant, 1.37×10^{-23} joules/degree
K	Stability coefficient
K_1	Impedance matching coefficient, = 1 for a perfect match
L_{ga}	Gate length in μm
L_g	Gate inductance
L_d	Drain inductance
L_s	Source inductance
$\mathcal{L}(f_m)$	Noise power density at the offset frequency
ℓ	Length of transmission line
L_m	Matching inductance
λ	DC parameter to adjust DC output resistance
λ_e	Wavelength
MESFET	Metal Semiconductor Field Effect Transistor
MMIC	Monolithic Microwave Integrated Circuits
$m(t)$	Amplitude modulation index
m_{in}	Coefficient representing input amplitude modulation
m_{out}	Coefficient representing output amplitude modulation
MG	Multiplication gain
n	Harmonic number and/or multiplication factor
n_i	Diode ideality factor
ω_H	High end of the frequency band
PHEMT	Pseudomorphic High Electron Mobility Transistor
P_{DC}	DC applied power
P_{out}	Output power
P_{DISS}	Power dissipated by the device

P_n	Output power at harmonic frequency n^{th}, $n = 0, 1, 2, 3 \ldots$
P_{nmax}	Maximum power at n^{th} harmonic
P_{dn}	Drain power at the nth harmonic
P_{osc}	Oscillator output power
P_{hose}	Harmonic oscillator output power
P_{of}	Drain oscillator power at the fundamental frequency
P_{oh}	Drain oscillator power at harmonic frequency
P_{av}	Average power
P_{ad}	Added power of an amplifier or multiplier
q	electron charge, 1.6021×10^{-19} C
$Q(V)$	Capacitor charge as a function of applied voltage
R_o	Radius of stability circle
$R_L(\omega)$	Load resistance at frequency ω
R_s	Source resistance
R_d	Drain resistance
R_g	Gate resistance
R_i	Depletion layer charging resistor
R_{ch}	Channel resistance
R_{Lopt}	Optimum drain load impedance
R_f	Feedback resistor
$R_D(\omega)$	Device resistance at frequency ω
T_{pp}	Coefficient for PM noise to PM noise conversion
T_{aa}	Coefficient for AM noise to AM noise conversion
T_{ap}	Coefficient for AM noise to PM noise conversion
T_{pa}	Coefficient for PM noise to AM noise conversion
T_a	Absolute temperature, in Kelvin
T	Period of sinusoidal voltage or current
2ϕ	Conduction angle in degrees
τ	represents the time delay in the gate-source voltage
$(\tau_0 + \tau_1)$	Conduction time
V_s	Magnitude of signal amplitude
V_{car}	Magnitude of the carrier voltage
V_c	Control voltage; voltage applied to capacitor C_{gs}
V_L	Load voltage
V_{SSB}	Amplitude of the single-sideband noise
V_{GS}	DC voltage across the gate-source diode

V_{GD}	DC voltage across the gate-drain diode
V_{DS}	DC voltage across the drain-to-source terminals
$V_{ds}(t)$	Total drain voltage $V_{ds}(t) = V_{D0} + V_{dn}\cos\omega t$
V_{D0}	Drain bias voltage
V_{dn}	Drain voltage at the nth harmonic
V_{GG}	Gate power supply voltage
V_{DD}	Drain power supply voltage
V_{DS0}	Drain voltage where A_0, A_1, A_2, A_3 are determined
$V_{gs}(t)$	Total gate voltage $V_{gs}(t) = V_{G0} + V_g\cos\omega t$
V_{G0}	Gate bias voltage
V_g	Gate voltage amplitude
V_P	Pinch-off voltage, $V_P = V_T + V_\phi$
V_T	Threshold voltage
V_ϕ	Built-in voltage
V_{gi}	Internal gate voltage
V_{di}	Internal drain voltage
V_k	Knee voltage
V_{21}	Fundamental frequency voltage at terminal 2
V_{22}	Second harmonic frequency voltage at terminal 2
V_{DM}	Maximum drain voltage
V_0	Voltage that describes the breakdown effect
$X_L(\omega)$	Load reactance at frequency ω
$X_D(\omega)$	Device reactance at frequency ω
W_g	Gate width in μm
Y_0	Transmission line admittance, default value = 20 mS
Z_0	Transmission line impedance, default value = 50 Ohm
$Z_L(\omega)$	Load impedance at angular frequency ω
$Z_D(\omega)$	Device impedance at angular frequency ω
Z_{insc}	Input impedance with short-circuited drain
Z_{inoc}	Input impedance with open-circuited drain

About the Author

Edmar Camargo was born in Paraná, Brazil on September 13, 1948. He was educated in São Paulo, Brazil, where he received his Electrical Engineering degree from Faculdade de Engenharia Industrial in 1972. He obtained his Master and Doctorate degrees in Electrical Engineering in 1977 and 1985, respectively from Universidade de São Paulo.

In 1973 he joined Philco-Ford to work on TV receivers, and in the next year, he joined the Laboratorio de Microeletronica at Universidade de São Paulo where he worked for 17 years. Besides teaching, he worked in the development of several microwave components for microwave communications for state companies and also for the industry. He took a leave of absence to work at Centre National d'Études des Telecommunications (CNET) in Lannion, France in 1977 and in 1982. He also worked as a consultant for Avantek and Fujitsu FCSI in 1991 and 1993, respectively. In 1993 he moved to Fujitsu to work on the design of power amplifiers and on MMIC design. In 1996 he moved to Hewlett-Packard Company where he worked on the design of mmWave modules, and in 1998, returned to Fujitsu where he is designing commercial MMICs.

Dr. Edmar Camargo is a senior member of the MTT Society, and a founding member of Sociedade Brasileira de Microondas e Optoeletronia (SBMO).

Index

The Artech House Microwave Library

Advanced Automated Smith Chart Software and User's Manual, Version 3.0, Leonard M. Schwab

Analysis, Design, and Applications of Fin Lines, Bharathi Bhat and Shiban K. Koul

Analysis Methods for Electromagnetic Wave Problems, Eikichi Yamashita, editor

C/NL2 for Windows: Linear and Nonlinear Microwave Circuit Analysis and Optimization, Software and User's Manual, Stephen A. Maas and Arthur Nichols

Computer-Aided Analysis, Modeling, and Design of Microwave Networks: The Wave Approach, Janusz A. Dobrowolski

Computer-Aided Analysis of Nonlinear Microwave Circuits, Paulo J. C. Rodrigues

Design of FET Frequency Multipliers and Harmonic Oscillators, Edmar Camargo

Designing Microwave Circuits by Exact Synthesis, Brian J. Minnis

Dielectric Materials and Applications, Arthur von Hippel, editor

Dielectrics and Waves, Arthur von Hippel

Electrical and Thermal Characterization of MESFETs, HEMTs, and HBTs, Robert Anholt

Frequency Synthesizer Design Toolkit Software and User's Manual, Version 1.0, James A. Crawford

Fundamentals of Distributed Amplification, Thomas T. Y. Wong

Generalized Filter Design by Computer Optimization, Djuradj Budimir

GSPICE for Windows, Sigcad Ltd.

HELENA: HEMT Electrical Properties and Noise Analysis Software and User's Manual, Henri Happy and Alain Cappy

HEMTs and HBTs: Devices, Fabrication, and Circuits, Fazal Ali, Aditya Gupta and Inder Bahl, editors

High-Power Microwaves, James Benford and John Swegle

LINPAR for Windows: Matrix Parameters for Multiconductor Transmission Lines, Software and User's Manual, Antonije Djordjevic, Miodrag B. Bazdar, Tapan K. Sarkar, Roger F. Harrington

Low-Angle Microwave Propagation: Physics and Modeling, Adolf Giger

MATCHNET: Microwave Matching Networks Synthesis, Stephen V. Sussman-Fort

Matrix Parameters for Multiconductor Transmission Lines: Software and User's Manual, A. R. Djordjevic, et al.

Microelectronic Reliability, Volume I: Reliability, Test, and Diagnostics, Edward B. Hakim, editor

Microstrip Lines and Slotlines, Second Edition, K. C. Gupta, Ramesh Garg, Inder Bahl, and Prakash Bhartia

Microwave and Millimeter-Wave Diode Frequency Multipliers, Marek T. Faber, Jerzy Chamiec, Miroslaw E. Adamski

Microwaves and Wireless Simplified, Thomas S. Laverghetta

Microwave Engineers' Handbook, Two Volumes, Theodore Saad, editor

Microwave Materials and Fabrication Techniques, Second Edition, Thomas S. Laverghetta

Microwave and Millimeter Wave Heterostructure Transistors and Applicatons, F. Ali, editor

Microwave and Millimeter Wave Phase Shifters, Volume I: Dielectric and Ferrite Phase Shifters, S. Koul, and B. Bhat

Microwave and Millimeter Wave Phase Shifters, Volume II: Semiconductor and Delay Line Phase Shifters, S. Koul and B. Bhat

Microwave Mixers, Second Edition, Stephen Maas

Microwave Transmission Design Data, Theodore Moreno

Microwave Transition Design, Jamal S. Izadian and Shahin M. Izadian

Microwave Transmission Line Couplers, J. A. G. Malherbe

Microwave Tubes, A. S. Gilmour, Jr.

Microwaves: Industrial, Scientific, and Medical Applications, J. Thuery

Microwaves Made Simple: Principles and Applicatons, Stephen W. Cheung, Frederick H. Levien, et al.

MMIC Design: GaAs FETs and HEMTs, Peter H. Ladbrooke

Modern GaAs Processing Techniques, Ralph Williams

Modern Microwave Measurements and Techniques, Thomas S. Laverghetta

Monolithic Microwave Integrated Circuits: Technology and Design, Ravender Goyal, et al.

MULTLIN for Windows: Circuit-Analysis Models for Multiconductor Transmssion Lines, Software and User's Manual, Antonije R. Djordjevic, Darko D. Cvetkovic, Goran M. Cujic, Tapan K. Sarkar, Miodrag B. Bazdar

Nonuniform Line Microstrip Directional Couplers, Sener Uysal

PC Filter: Electronic Filter Design Software and User's Guide, Michael G. Ellis, Sr.

PLL: Linear Phase-Locked Loop Control Systems Analysis Software and User's Manual, Eric L. Unruh

The RF and Microwave Circuit Design Cookbook, Stephen A. Maas

RF Design Guide: Systems, Circuits, and Equations, Peter Vizmuller

Scattering Parameters of Microwave Networks with Multiconductor Transmission Lines: Software & User's Manual, A. R. Djordjevic, et al.

Solid-State Microwave Power Oscillator Design, Eric Holzman and Ralston Robertson

Terrestrial Digital Microwave Communications, Ferdo Ivanek, et al.

Transmission Line Design Handbook, Brian C. Waddell

TRAVIS Pro: Transmission Line Visualization Software and User's Manual, Professional Version, Robert G. Kaires and Barton T. Hickman

TRAVIS Student: Transmission Line Visualization Software and User's Manual, Student Version, Robert G. Kaires and Barton T. Hickman

Yield and Reliability in Microwave Circuit and System Design, Michael Meehan and John Purviance

For further information on these and other Artech House titles, including previously considered out-of-print books now available through our In-Print-Forever™ (IPF™) program, contact:

Artech House
685 Canton Street
Norwood, MA 02062
781-769-9750
Fax: 781-769-6334
Telex: 951-659
e-mail: artech@artech-house.com

Artech House
Portland House, Stag Place
London SW1E 5XA England
+44 (0) 171-973-8077
Fax: +44 (0) 171-630-0166
Telex: 951-659
e-mail: artech-uk@artech-house.com

Find us on the World Wide Web at: www.artech-house.com